Studies in Computational Intelligence

Volume 1162

Series Editor

Janusz Kacprzyk, Polish Academy of Sciences, Warsaw, Poland

The series "Studies in Computational Intelligence" (SCI) publishes new developments and advances in the various areas of computational intelligence—quickly and with a high quality. The intent is to cover the theory, applications, and design methods of computational intelligence, as embedded in the fields of engineering, computer science, physics and life sciences, as well as the methodologies behind them. The series contains monographs, lecture notes and edited volumes in computational intelligence spanning the areas of neural networks, connectionist systems, genetic algorithms, evolutionary computation, artificial intelligence, cellular automata, self-organizing systems, soft computing, fuzzy systems, and hybrid intelligent systems. Of particular value to both the contributors and the readership are the short publication timeframe and the world-wide distribution, which enable both wide and rapid dissemination of research output.

Indexed by SCOPUS, DBLP, WTI AG (Switzerland), zbMATH, SCImago.

All books published in the series are submitted for consideration in Web of Science.

Anirudha Joshi · Gavin R. Sim
Editors

Proceedings of the 14th Indian Conference on Human-Computer Interaction

IndiaHCI'23

India HCI **2023**

Springer

Editors
Anirudha Joshi
IDC School of Design
Indian Institute of Technology Bombay
Mumbai, Maharashtra, India

Gavin R. Sim
School of Engineering and Computing
University of Central Lancashire
Preston, Lancashire, UK

ISSN 1860-949X ISSN 1860-9503 (electronic)
Studies in Computational Intelligence
ISBN 978-981-97-4334-6 ISBN 978-981-97-4335-3 (eBook)
https://doi.org/10.1007/978-981-97-4335-3

This Springer imprint is published by the registered company Springer Nature Singapore Pte Ltd.
The registered company address is: 152 Beach Road, #21-01/04 Gateway East, Singapore 189721, Singapore

Preface

IndiaHCI series of conferences are Indian conferences on Human-Computer Interaction. It was first organized in 2004 and has been regularly organized since 2010. Earlier conferences were organized by groups of volunteers. HCI Professionals Association of India was formed as a not-for-profit organization in 2013, which has since taken over the financial responsibility of the conference. Previous conferences were held in different cities in India, namely Mumbai, Bangalore, Pune, Delhi, Guwahati, and Hyderabad.

IndiaHCI 2023 conference was the 14th Indian Conference on Human-Computer Interaction. It was held at UPES, Dehradun, from November 23–25, 2023.

Review Process:

(1) Track—Full Papers:

Each paper underwent a rigorous review process managed by the two Paper Chairs (Program Committee Chairs). Each paper was assigned two Associate Chairs and two external reviewers, all of whom were experts in the field of HCI. The second Associate Chair (2AC) and the external reviewers provided blind reviews. The first Associate Chair (1AC) provided a meta-review. Approximately, 3 to 4 papers were assigned to each reviewer for assessment. Authors of the selected papers received the initial reviews for their work and had the opportunity to submit a 500-word rebuttal to address questions or clarify any misunderstandings raised by the reviewers. The final decision regarding accepting or rejecting a paper was reached during a virtual Program Committee (PC) meeting where the Paper Chairs, the 1AC, and the 2AC discussed the papers. Conflicts were managed during the PC meeting. High-quality papers were accepted directly. Papers showing promise but requiring changes were shepherded and accepted only if the Shepherd agreed. All acceptances were conditional, contingent on potential changes suggested by the papers committee for the final camera-ready draft. The primary author of each accepted paper received detailed instructions on how to submit a final, publication-ready version of the paper. Formal acceptance into IndiaHCI was granted once authors revised their submissions and submitted a final draft for approval by the Program Committee.

(2) **Track—Work-in-Progress Papers:**

Some of the papers that were submitted to the full paper track were selected for the Work-In-Progress track. A "Work-in-Progress Paper" is a new category of acceptance. This means that the Program Committee likes the novelty in the work but does not consider that the research is ready yet to be accepted as a full paper. As a Work-in-Progress Paper, the authors were given a chance to present their work at the conference. The review process of Work-In-Progress papers is the same as the full papers track.

There were 23 submissions for the papers track, of which five were accepted as full papers and three as Work-in-Progress papers.

(3) **Track—Arts Track Papers (Short Papers):**

The arts track, hosted for the first time at IndiaHCI 2023, aimed to cultivate a community of artists and researchers who explored the intersectional space of art, design, technology, and HCI. Artworks were in various forms including art installations, performances, storytelling experiences, and experimental artworks using HCI, complemented by a short paper submission.

The arts track chairs and a panel of domain experts reviewed submissions through a single-blind peer-reviewed process. Each paper received two to three reviews. There were 17 submissions for the arts track, of which four were accepted for publication.

Mumbai, India
Preston, UK
February 2024

Anirudha Joshi
Gavin R. Sim

Contents

Full Papers

Framework for Conducting Maturity Evaluation of Teleconsultation Platforms: A COVID-19 Motivated Project from India ... 3
Ayushi Tandon, Uma Nambiar, Madhava Sai Sivapuram, and Avinash Kumar

***"So, Should I Walk Today or Not?"* Understanding Concerns and Queries on Health and Fitness Among Indian Older Adults** ... 23
Pallavi Rao Gadahad and Anirudha Joshi

Design and Evaluation of a Password Diversifier Tool ... 51
Rekha Pathak, Anjali Aggarwal, R. Aiswarya, Jayasree Raveendran, Manish Shukla, Vijayanand Banahatti, and Sachin Lodha

Graphical Passwords for Emergent Users: A Four-Day Recall Comparative Study on PIN, Passfaces and Celebrities ... 75
Asif P Zuha, Anirudha Joshi, Sanika Deshpande, Anniket Mohalik, Rohit Naik, and Vijayanand Banahatti

Is JOMO the Answer to FOMO? Collaborative Digital Detox from Social Media for Adolescents ... 99
Anjali Mishra, Sanchita Jagdishkumar Bagre, Wricha Mishra, and Sumita Sharma

Work in Progress Papers

Revival of the Traditional Gond Art of Madhya Pradesh Through Modern Technologies of the Digital Platforms and UX-UI ... 121
Harsh Raj Gond and Ankita Roy

Touchtalk: Enhancing Messaging in Smartphones for Visually Impaired 133
Bindi Shah and Mrunmai Abhyankar

Navigational Integration: Enhancing User Experience Through Illustration-Driven Wayfinding in a Multi-storied Building in Real and Augmented Environments 149
R. Gokul and Ankita Roy

Arts Track Papers (Short Papers)

Belonging: Exploring Audio-Visual Cues in Real-Time 6DoF VR Narrative 171
Jayesh Pillai, Priyanshu Thakur, and Tushar Vaid

Whispers of the Konkan Guardian: Interactive Storytelling Through Digital Game Design 179
Shreyas Vernekar and Jayesh Pillai

***KaavadBits*: Interactive Art Installation for a New Narrative Experience** 189
Saumik Shashwat, Aditya Padmagirwar, Shivoy Arora, and Anmol Srivastava

Jail Diaries: An Interactive Pop-Up Book Art Installation 197
Biswasmita Mishra, Aishiki Bhattacharya, Ankush Gupta, Siddharth Gupta, and Anmol Srivastava

About the Editors

Anirudha Joshi works in the field of Human-Computer Interaction (HCI) design. He works in two research areas—interaction design for emergent users in developing countries and integrating HCI with software engineering. Anirudha specializes in user studies, interaction design, and usability evaluations. He has worked in diverse domains, including healthcare, banking, education, literacy, text input, industrial equipment, and FMCG. His work has been on platforms such as desktops, the web, mobile phones, IVRs, ATMs, and custom hardware.

Gavin R. Sim research lies in the fields of human-computer interaction and child-computer interaction. He focuses on designing and evaluating technologies in a wide range of contexts, including gaming, education, and health. He also uses various methods including Participatory design, survey tools, and technologies including eye tracking, to improve end users' experiences.

Full Papers

Framework for Conducting Maturity Evaluation of Teleconsultation Platforms: A COVID-19 Motivated Project from India

Ayushi Tandon, Uma Nambiar, Madhava Sai Sivapuram, and Avinash Kumar

Abstract We conducted online evaluations of 58 teleconsultation platforms to identify the expectations and feature requirements of multiple stakeholders, including doctors, hospital administrators, hospital IT managers and health technology platform vendors. We then analyzed evaluation summaries and propose a maturity evaluation framework; a consensus development process was employed and experts from all participating organizations in the alliance were engaged. The resultant maturity framework is descriptive in nature, with three levels presented in a clear and concise manner. We recommend to view maturity as a dynamic and ongoing process. Patient-centric features were found to be desirable during evaluations, but ultimately took a backseat to medical and institutional logistics in terms of relevance for maturity evaluation. Compatibility of teleconsultation platforms with existing infrastructure and clinical practices was deemed important by all experts due to concerns about adoption and seamless transition to the new platform. This framework can serve as a roadmap for teleconsultation platform developers and could also be used to assess the teleconsultation platform maturity within a country, or organization.

Keywords Teleconsultation · Maturity framework · Platform design · Tele health · Teleconsultation evaluation framework · Teleconsultation in India

---

A. Tandon (✉)
School of Management, Mahindra University, Hyderabad, India
e-mail: ayushi.tandon@mahindrauniversity.edu.in

U. Nambiar · A. Kumar
Digital Health India Association, Bengaluru, India

M. S. Sivapuram
Dr Pinnamaneni Siddhartha Institute of Medical Sciences and Research Foundation, Chinna-Avutapalli, India

A. Joshi and G. R. Sim (eds.), *Proceedings of the 14th Indian Conference on Human-Computer Interaction*, Studies in Computational Intelligence 1162,
https://doi.org/10.1007/978-981-97-4335-3_1

1 Introduction

Over the past three decades, there has been a notable proliferation of research articles examining the implications, antecedents, and consequences associated with the utilization of telehealth platforms. These investigations have spanned both developed and developing countries. However, there has been a noticeable absence of guidance on selecting or developing an appropriate telehealth platform. Hospital administrators, physicians, and various stakeholders of telehealth ecosystem face a multitude of platform choices, yet they encounter challenges in evaluating these options effectively, often lacking the requisite tools to determine whether the chosen platforms align adequately with their requirements. Thus, uncertainty to assess their offerings has been felt by both providers and users of these platforms, underlining the need for comprehensive evaluation framework for telehealth platforms [11, 29]. In the field of telehealth, teleconsultations play a crucial role as they serve as the fundamental framework for linking patients with remote healthcare services and maintaining the continuity of care. Moreover, during the initial stages of the COVID-19 pandemic, efforts to reduce in-person healthcare visits, especially in hospital outpatient departments worldwide, resulted in the widespread adoption of teleconsultations as the preferred approach for rapid clinical assessment and delivering healthcare remotely. This shift was embraced not only by patients and healthcare professionals but also by hospitals [12, 36, 45]. Simultaneously, legislative changes in favor of telehealth delivery stimulated health technology entrepreneurs to introduce their products for telehealth delivery, most common being teleconsultation platform, in both developed and emerging nations [41, 64, 68]. Our examination of existing research shows that many studies have looked into the effects and use of teleconsultation services, both before and during the initial years of the COVID-19 pandemic. But, no comprehensive framework exists that could be applied to evaluate the maturity of current teleconsultation platforms and to advance their development. This gap can be in part attributed to the simplistic perception of teleconsultation platform as remote healthcare delivery tool with features such as video or image transfer, without considering the social or organizational aspects that influence the utilization of such platforms. While it is noteworthy that in resource-constrained and diverse regions, like many countries in the Global South, social or organizational aspects are further complicated by disparities in healthcare provider-patient ratios, nascent digital health policies, and social and cultural contexts that mediate healthcare utilization [13]. This paper is situated at the convergence of these two distinct healthcare research contexts. On one hand, there is a need to develop framework for evaluating the teleconsultation platforms. On the other hand, it is imperative to account for the complexities within healthcare ecosystems at the social, organizational, and national levels. To this end, an exploratory study on teleconsultation was conducted in India through a research alliance involving five organizations committed to enhancing healthcare delivery through information and technology. These organizations include the National Resource Centre for EHR Standards, the College of Healthcare Information Management, the Consortium of Accredited Healthcare Organizations, the

Healthcare Information and Management Systems Society (India), and Digital Health India. Representatives from these organizations were nominated to form an expert working group comprising 12 members. Based on the work this research alliance this paper attempts to answer, *how should a tele-consultation platform be evaluated for its maturity that account for complexities of remote healthcare delivery in India?* Maturity here is not conceptualized as the success of one specific technological system; instead, it represents progress in aligning features with the requirements of the teleconsultation ecosystem, which includes healthcare providers, product developers, hospital administrators, and patients. Furthermore, we aim to approach this assessment in a manner that departs from the conventional standards tied to clinical users' specific services typically used to evaluate telehealth systems.

2 Background

There are three phases to the current study's execution. The first phase involved survey of previous literature. We reviewed the studies obtained from three databases, namely Scopus, Web of Science, and Google Scholar. The selection of included studies for this paper followed the Preferred Reporting Items for Systematic Reviews and Meta-Analyses (PRISMA) guidelines, focusing on the timeline between 2010 and 2020. To enhance the relevancy of search results, the key search terms used were "teleconsultation evaluation," "teleconsultation maturity," "teleconsultation framework," "Telehealth framework," "Telehealth maturity," "Telemedicine framework," and "Telemedicine maturity." These key search terms were combined using Boolean operators (OR/AND). The initial screening was done on the basis of titles and keywords and it involved selecting studies that evaluated telemedicine in general, with a specific focus on teleconsultation. In addition to peer-reviewed published studies with high indexing, a literature search was conducted to incorporate popular frameworks and studies among practitioners. Relevant academic articles and materials related to these frameworks were included in the initial database. The next screening of abstracts was performed by the third and fourth authors, who identified papers that provided a systematic evaluation of one or more aspects of teleconsultation. This process resulted in the identification of 132 studies for in-depth reading. In order to synthesize the literature, a total of 132 selected studies were read and categorized based on their key focus, which included both single element and synthesis of diverse elements, following Maeder and Poultney's [48] approach. Only studies that specifically evaluated teleconsultation were included. We also documented the stakeholders involved as focal participants and classified the theoretical approaches used in each study according to Currie's 2005 [22] classification, as presented in Table 1. Moreover, we analyzed the types of outcomes investigated as a result of utilizing teleconsultation platforms, as well as the research methodologies employed by the researchers in selected articles. Our analysis referred to the works of Coiera [20] in 2003 and Andargoli et al. [4] in 2017 to understand the approaches used in studying these outcomes. After an extensive screening process, we have included

21 papers that met our inclusion criteria and provide a comprehensive overview of the teleconsultation platform evaluation. These papers are listed in Table 1, offering valuable insights into the evaluation of teleconsultation.

The findings from included studies highlight that in the prior research frequently cited evaluation frameworks include the Model for Assessment of Telemedicine (MAST) [42], RE-AIM model [33], Technology Acceptance Model (TAM) [39], and Donabedian's Quality of Care Model (QOCM) [10]. Outcome evaluation studies on teleconsultation platforms are often conducted at a behavioral level, with the Technology Acceptance Model (TAM). The RE-AIM and Modified Delphi Method recommend consulting experts in the development of evaluation tools, and have been utilized to assess the usability, utility, and integration of teleconsultation platforms with other healthcare IT systems [3]. In addition, frameworks such as the Capability Maturity Model and the Telemedicine Maturity Model (TMM) in studies on the feasibility evaluation at the organizational level [65]. Table 2 summarize the frameworks mentioned. Prior research (Table 1) has largely focused on the evaluation of teleconsultation platforms from the perspective of a single user, rather than adopting a more diverse multi user approach. Key outcomes of interest to researchers include the low cost and advanced clinical information technology integration, as well as the technical suitability and usability of teleconsultation platforms for clinical and administrative purposes [24, 65]. Ease of use, clinical utility, and economic accessibility are commonly considered parameters in the evaluation [19, 24]. Overall, we find that studies conducted on teleconsultations tended to neglect the situational factors influenced by the complex dynamics within healthcare systems, spanning social, organizational, and national levels. While scholars that investigated health systems in global south have shown that intricate dynamics of health technology adoption were further compounded by disparities in the ratios of healthcare providers to patients, the evolving landscape of digital health policies, and the social and cultural factors that shape healthcare utilization [14, 16]. Therefore, it is imperative to undertake research utilizing suitable methodologies that engage a variety of domain experts [34, 51, 60] to identify the necessary features for successful teleconsultations. This is the gap that our study partially addressees.

3 Method

3.1 Data Collection and Analysis

The phase two of study was carried out between May 2020 and October 2021. We conducted activity walkthrough evaluation by experts for teleconsultation platforms that expressed interest in being part of the study [28, 32, 44, 49]. The experts comprised of professionals with experience in healthcare delivery, organizational and administrative policy, telemedicine research, senior managers, and health-care consultants (Table 3). Teleconsultation platform vendors were recruited to participate

Table 1 Categorization of teleconsultation evaluation studies

Article reference	Single or diverse [48]	For whom	Currie's classification [22]	What (content)	How
Yan et al. [67]	Diverse	Doctors	Behavioral	Use as outcome	Qualitative enquiry, case study
Alkmim et al. [1]	Single	Doctors	Generic	Clinical usability	Quantitative summative
Alkmim et al. [2]	Single	Healthcare workers	Generic	Ease of use	Quantitative survey
Petcu et al. [55]	Single	Patients	Behavioral	Clinical usability	Quantitative survey
Esterle and Mathieu-Fritz [27]	Single	Patients and care givers	Social relationship	Clinical outcome	Qualitative observation and interview
Maarop and Win [47]	Single	Healthcare providers	Behavioral	Clinical usability	Quantitative survey and qualitative interviews
Grenier et al. [35]	Single	Patients	Behavioral	Clinical outcome	Quantitative survey
Lu et al. [46]	Diverse	Doctors	System development	Usability and organizational fit	Review and quantitative survey
Gallerini et al. [31]	Single	Patients	Generic	Usability and organizational fit	Qualitative observation-based enquiry
De la Torre Díez et al. [63]	Diverse	Patients	System development	Usability and organizational fit	Quantitative survey
Czekierda et al. [23]	Diverse	Doctors	System development	Clinical usability	Qualitative case study
Park et al. [53]	Single	Patients	System development	Clinical usability	Quantitative survey
Paul and McDaniel [54]	Single	Clinicians, administrators and IT professionals	Behavioral	Adoption outcome	Qualitative case study
Barrett [7]	Single	Nurses	Social relationship	Clinical usability	Qualitative case study
Rimner et al. [57]	Single	Patient	Generic	Ease of use	Quantitative survey
Gackowski et al. [30]	Single	Doctors	System development	Clinical usability	Quantitative experiment
Toledo et al. [62]	Single	Patient	Behavioral	Clinical outcome	Quantitative survey
Mahnke et al. [50]	Single	Doctors	System development	Usability and organizational fit	Qualitative observation
Li et al. [43]	Diverse	Doctors and patients	System development	Clinical usability	Quantitative and qualitative
Viers et al. [66]	Single	Patients and doctors	Behavioral	Clinical usability	Quantitative experiment
Tchero et al. [61]	Single	Patients	Generic	Clinical outcome	Quantitative experiment
This study	Diverse	Patient, doctors, CIOs, health administrators	System development	Features-level maturity of technology platform	Qualitative

Table 2 Summary of popular evaluation frameworks used in teleconsultation studies

Framework	Focus area
Model for assessment of telemedicine (MAST)	Evaluating telemedicine outcomes across various domains, including access to care, cost, experience, and effectiveness
RE-AIM model	Utilized to assess the usability, utility, and integration of teleconsultation platforms with other healthcare IT systems
Technology Acceptance Model (TAM)	Commonly used at a behavioral level for intention to use and adoption. Derivatives of TAM are also used in these studies
Capability Maturity Model, Telemedicine Maturity Model (TMM) and Telehealth Maturity Model (THMM)	Feasibility evaluation at the organizational level. Used to examine factors such as technology infrastructure, regulatory environment, and funding sources for the implementation of teleconsultation platforms within an organization
This study	Proposes a framework to assess the maturity of teleconsultation platforms in terms of their ability to offer the progression of features aligned with the demands of the teleconsultation ecosystem

in the study via survey-based recruitment using various social media platforms such as Facebook, LinkedIn. The walkthrough method for conducting a usability test comprises two main stages: preparation and execution. For first stage, to ensure the accuracy and comprehensiveness of identified use cases and scenarios (see Table 4 for comprehensive list) nominated experts from various alliance organizations were involved in the process. Moving on to the execution stage, it involved a sequential walkthrough of actions and the recording of feedback by the experts. 58 vendors participated and each vendor was given a 45 min online slot to demonstrate the workflow of their teleconsultation platform, starting from basic activity and then experts asked specific activity demonstrations. Following each demonstration, the expert evaluators discussed and provided subjective notes on how well it addressed scenarios enquired. These notes were combined by the first, third and fourth authors to create a comprehensive evaluation summary for each teleconsultation platform. The first and second authors attended all demonstrations and evaluation discussion sessions. In the third phase of the study, we adopted a consensus development process [38] commonly used to develop guidelines in the healthcare field by synthesizing available information and expert opinions [21]. Inputs from experts Profile in Table 4 were synthesized by first and second author and presented in the form a framework to take their qualitative inputs. The first version of framework had six levels and three feature dimensions (mandatory features, security features, and aspirational features). In total we held five online meetings with experts before finalizing the framework via voting on feature categorization and framework levels [40]. All discussions were

Table 3 Profile of nominated experts that were part of evaluation and framework development working group

Domain	Professional role	Experience (years)
Medical policies and healthcare administration	Strategic healthcare consultant	20+
Health informatics and user engagement	Academic research	4
Standards and interpretability in digital health	Director product	20+
Healthcare services administration	Lead product	10+
Standards and interpretability in digital health	Joint director product	15+
Digital products development	Lead platforms of digital products	20+
Digital transformation healthcare	Lead product	10+
Medical care	Senior physician	15+
Digital transformation healthcare	Strategic healthcare consultant	20+
Information security	Director	10+
Digital transformation healthcare	Director	20+
Medical care	MBBS	5+

carefully documented, and the framework was subsequently modified to address any concerns raised by the experts. We also conducted an online workshop about framework with teleconsultation vendors that participated in evaluation exercise for their input. The teleconsultation maturity evaluation framework resulting from this process is presented in the next section of the paper.

3.2 *Positionality*

It's important to note that all authors possess relative social and economic capital in Indian society that may distance them from the marginalization experienced by patients who rely on technology-mediated teleconsultation. Nonetheless, drawing from our collective experiences of studying and participating in healthcare delivery among underserved populations in rural and urban India, we were motivated to emphasize patient needs when collaborating with other experts during consensus building on framework. We sought to highlight features that could potentially contribute to more equitable and inclusive teleconsultation offerings. While this goal emerged as a long-term objective in the final framework, it is mentioned here to transparently convey our positionality as we proceed to discuss the findings and framework.

Table 4 Comprehensive list of features and scenarios enquired by the experts during evaluations

Basic activities (8 vendors demonstrated only these activities)	Advance activities (39 platforms demonstrated some features for listed activities)	Automation activities (11 platforms had features to support automated teleconsultation workflow)
Onboarding of patients	Doctor and patient profile management	Pre-appointment triage support
Onboarding of doctors	Dashboard layout for doctors	Risk assessment of patient
Modes of interaction between doctor and patients	Templates and customizations available for doctor	Digital assistance for medical record keeping
Clinical documentation features for doctors	Facility for referral and multi-doctor team consultation	Standardizations (security, databases, data storage, data exchange)
Prescription format	Consent modalities	Patient portal and facilities for patient health record
Features for patient to share documents	Provision for record keeping	Health education and counselling support
Alternative to smart phone ownership for patients	Feature for caregivers of elderly and adolescent	Assistance to patient in following the care plan
Consent types for teleconsultation session	Interoperability and integration with hospital information system	IoT integration
	Drug data base integration	Features for secondary remote care
	Clinical decision assistance	Warning system in line with guidelines, standards and protocols
	Administrative support features	Patient evaluation and feedback
	Provision for data security Multilingual suppor	Integration with third-party services

4 Findings

We analyzed the summaries of demonstrations generated by the experts during the evaluation process. The platforms are divided into three groups based on the overall review given by the experts. We compiled a list of scenarios and situations that were considered by the experts during their evaluations, and documented those platforms (see Table 4 and Fig. 1). Experts reached consensus on following points: (1) Doctors are concerned about the amount of time they need to spend on documentation and setup tasks, so platforms that are easy to use and navigate would be preferred. For doctors, platform should support necessary medical documentation and legally acceptable modes of communication in order to ensure continuity of care; (2) For patients, clear communication is essential to understand information and instructions provided by doctors, both verbally and through prescribed treatment or medication. It is also valuable for the platform to provide secure connections and protect patient confidentiality, as well as allow for the exchange of information with caregivers, other healthcare specialists, and support staff; (3) From the perspective of hospital administration, it should support regional and national healthcare approaches and provide data/reports for accreditations; (4) IT team expects the chosen platform to be easy to maintain and compliant with all relevant guidelines and laws. Experts agreed on the three-tier format of the framework with patient-centric features at top and standard compliant platform at middle tier (see Fig. 1). The above considerations provided structure to maturity evaluation framework. It is stratified into three levels (minimal, comprehensive and smart assist). Each level is further represented by three dimensions (interaction and communication, documentation, compliance) based on commonalities in features and product attributes and product can have varying maturity across three levels. Generally, platforms can offer a range of features spanning from a "minimal" layer to increasingly comprehensive features, ultimately culminating in the "smart assist" category. These features encompass three distinct domains: interaction and communication, documentation and data management, and compliance. However, it is important to note that in practice, there may be some degree of overlap and iteration between these layers across domains. For example, a product might incorporate smart assist features in the realm of interaction and communication while providing only minimal-level features for compliance, documentation, and data management. Summary of maturity framework across three levels is provided in Table 5. The maturity framework is descriptive in nature as we provide levels in simple and textual manner [65]. The minimal level of the framework is focused on providing the bare minimum of features or functionality needed to perform the teleconsultation. The comprehensive level builds upon the foundation of the framework, i.e., minimal level and adds features or functionality that help to ensure clinical quality and effectiveness of teleconsultation activity. Platforms at this level provide extensive features to help users meet their clinical and administrative objectives. The comprehensive level is thorough and include features for ensuring teleconsultation and its documentation comparable to physical consultation visit. The smart assist level of teleconsultation platforms as per the framework are focused on providing intelligent and compliant assistance to help users perform their tasks. The features are combined under this category such that the teleconsultation platform provides advanced and sophisticated assistance that is aided by intelligent systems or tools, and that is designed to be both effective and compliant with relevant rules or regulations. The three dimensions across which features are grouped in the proposed framework are elaborated next.

4.1 Interaction and Communication

Interaction and communication are essential capabilities of teleconsultation platforms, include enabling remote connections between doctors and patients. Experts

Table 5 Summary of teleconsultation maturity evaluation framework

Level	Description	Key capabilities summary
Minimal	Providing the bare minimum of features or functionalities	Provides multiple remote interaction modality (audio, video)
		Features for registering patients
		Supports asynchronous communication (email, SMS for prescription sharing)
		Payment management
		Scheduling and calendar management
Comprehensive	Features or functionalities that help to ensure quality and effectiveness of teleconsultation activity comparable to physical consultation visit	Synchronous communication (upload and download of document, image sharing)
		Advance interaction support (such as, in call chat, drawing pad, customizable information brochure, and private note taking)
		Compatibility with basic consultation workflow and supports specialty specific customization
		Platform facilitated proforma for assessment and treatment plan documentation and communicating documented details
		Referral management
		Identification and authentication features
		Provision of role-based login
		Onboarding training
		Consent management
		Data management including data provenance
		Integration with third party applications for interaction and communication
		Clinically relevant database integration (such as drug, diagnostic, hospital held electronic patient records)
Smart assist	Platform provides advanced assistance that is aided by intelligent systems or tools, and is designed to facilitate adherence with relevant rules or regulations	Consent management system adaptable for functionally illiterate patient, stigmatized clinical condition, etc.
		Multilingual and translation support wherever required by patient
		Interaction and communication support features for caregivers, minors, etc.
		Patient portal and provision for maintaining longitudinal health record
		Customizable alerts and reminders
		Recommendations on clinical assessment and treatment guidance based on latest research
		IoT integration
		Rule based optimization of varied features by applying learning algorithms
		Personalized training and grievance redressal
		Data lifecycle management

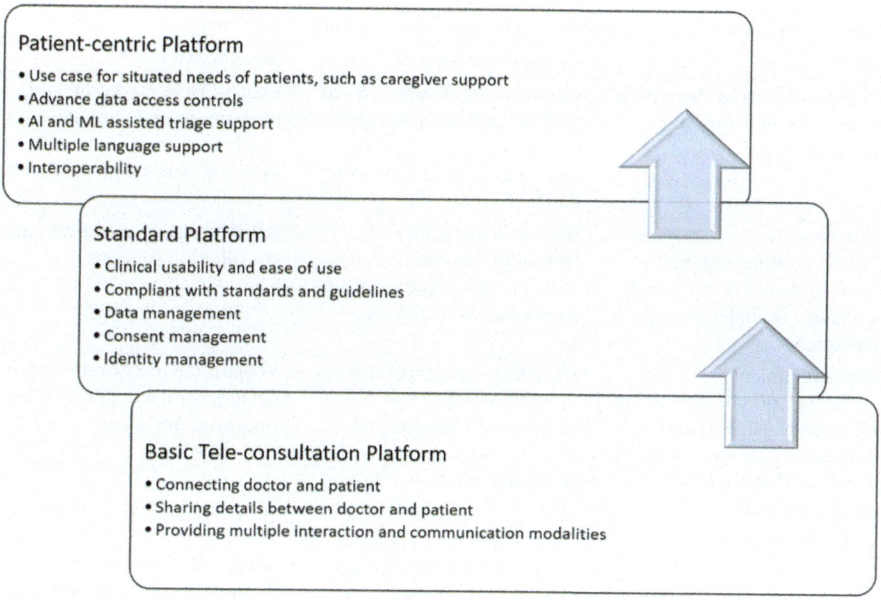

Fig. 1 Layered representation of maturity of teleconsultation platforms

emphasized the importance of platform features that allow patients to transmit their symptoms and signs to physicians remotely, including synchronous (e.g. audio or video consultations) and asynchronous (fax or email) exchange of dialogue and documents such as prescriptions. Experts have also included the features for identity management and administrative tasks on payment and onboarding as part of interaction and communication dimension. We have included comparison of all three levels across this dimension in Table 6.

4.2 *Documentation and Data Management*

Proper documentation of clinical interactions during teleconsultation sessions is essential. To this end, experts agreed that teleconsultation platforms should have a feature for documenting assessment and treatment plan. Platforms are expected to maintain a record of interactions that take place using their features, while doctors using platforms with minimal documentation features are responsible for any interactions outside the platform. As a result, maintaining longitudinal records of doctor-patient interactions on basic teleconsultation platforms may not be possible. Experts agreed that doctors using minimal feature platforms may need to rely on patients or their care givers to provide information on previous medical history, symptoms, and vital measurements in order to properly diagnose patients during remote consultations. Similarly, private note-taking features may be helpful for doctors consulting

Table 6 Interaction and communication features across three levels of maturity

Minimal level	Comprehensive level	Smart assist level
Enable patients to share their symptoms and signs to physicians remotely	Key features include screen sharing, and other features to enable doctors to explain assessment and diagnosis to patients	Flexibility in modes of interaction, including the ability to switch to audio if there are issues with bandwidth
Synchronous (e.g. audio or video consultations) and asynchronous (fax or email) exchange of dialogue and documents	Identity management including verifying, authenticating, and recording the identities of all users	Allow multiple parties to join consultation sessions
Support clinical documentation of interactions between physicians and patients, including verbal, visual, and text-based communication	Including digital identities or digitalized version of recognized identifiers of doctors and patients in communication	Support the use of remote devices for data collection and improved diagnosis
Features for sharing images of hand-written prescriptions	Easy appointment booking and multiple payment options	Integration with databases and advanced identification methods to verify doctors' credentials and establish the identity of patients and caregivers
Integration with WhatsApp messaging, and SMS	Cancellation and refund management	Smart recommendations, guidance in choosing a doctor during appointment
For certain situations features for remote teleconsultation centers	Real time appointment management	Advanced analytics, and integration of medication ordering and tracking
Management of doctor's availability	Tools for onboarding training for clinicians and patients	Application of rule-based learning to optimize the user interface and provide machine learning-based guidance to user
Setting prices and handle payments online		Inclusion of language translation
		Queue management

with patients on sensitive issues, but can also create challenges for advanced record search and retrieval. Thus, capabilities related to maintaining electronic health record and electronic patient records, along with patient portal etc. are grouped under this dimension. We have included comparison of all three levels across this dimension in Table 7.

Table 7 Documentation and data management features across three levels of maturity

Minimal level	Comprehensive level	Smart assist level
Feature for audio/video recording interactions between doctors and patients	Support the documentation within the teleconsultation workflow	Clinical decision support and integration with clinical and drug databases can help reduce the risk of errors in prescription
Flexibility to communicate with patients in multiple ways, such as sharing pictures and emailing prescriptions through the platform's mail server or external platforms	Template for documenting prescription as per guidelines (such as valid patient identity, generic name of the drug, dosage and formulation, specify frequency and route of administration, and the signature of the doctor)	Robust data management and data exchange capabilities that assist physicians in adhering to clinical rules and guidelines
Template for documenting perceptions, care plans	Private note-taking feature	Allow customization of the prescription template and teleconsultation workflow according to specialty or consulting style
Facility to record verbal consent as required by guidelines	Provide detailed, specialty-specific documentation template	Data base integration capabilities for alerting on drug allergies and doses
	Features for the retrieval of past records	Feature for referrals
	Customizable database to alert on medication errors and restricted drug prescription	Facilitate documentation in line with the goal of digital health records i.e. to have a single, longitudinal electronic record for each patient
	Plotting and charting of vital signs, and warnings and recommendations based on pre-programmed rules	Natural language processing, and voice typing capabilities
	Customizable permissions for document sharing	
	Maintain communication logs	

4.3 Compliance

These requirements may vary depending on national guidelines and policies. For instance, in India, it is important for the platform to identify the patient and the type of consultation, whether it is a first-time or follow-up visit to ensure that consent is implicit in the first scenario and executed explicitly between the telemedicine platform and the patient in the second scenario. It is important for the platform to facilitate a secure connection for teleconsultations and allow for flexibility in the use of third-party applications. Experts have also included the features for consent management and health data exchange as part of compliance dimension. We have included comparison of all three levels across this dimension in Table 8.

5 Discussion

The maturity evaluation framework proposed in this paper lists progression of features of a teleconsultation platform that are aligned with the teleconsultation ecosystem, encompassing healthcare providers, product developers, hospital administrators, and patients. Although, this framework departs from conventional clinical user and service-specific standards typically used to evaluate telehealth systems. But, we find that in the framework doctor-specific features align with previous research on the clinical needs of doctors during different consultations and specialty-specific consultations [18, 59, 67]. It also highlights the importance of features in teleconsultation platforms for identifying and communicating with patients, as well as for maintaining legally acceptable records that patients can understand and act upon. The features in comprehensive and smart assist level are also relevant for ensuring that patients trust the teleconsultation process, understand information and instructions provided by doctors, and have the option to switch to in-person care, if needed. Previous research has shown that assessing teleconsultation platforms features for both doctor and patient needs can significantly improve outcomes of post-implementation user satisfaction studies [25]. This goes beyond just reducing the time spent during medical interviews [26]. It is important to note that certain patient-centric features, as listed in Table 5, were considered important only during evaluations, as experts discussed the needs of patients with low income and living in remote areas [9, 58]. However, during the consensus development process for the maturity framework, these features lost their relevance for comprehensive teleconsultation platforms. Although patient-centric features were seen as ideal and desirable, it became apparent that "medical care logics" or "professional logics" and "managerial logics" took precedence over "patient-centered logics" [5, 6] during the process of abstraction. As a result, these patient-centric features were placed at the upper end of the maturity model (smart assist). This finding is also consistent with previous research on the limited utilization of patient-centric features and may explain why doctors resist the use of teleconsultation platforms, as they may disrupt their routines and not align with their "logic of care" [56]. In addition, many patients feel a disruption in their relationships with their doctors when interactions are mediated by technology platforms and tend to return to in-person consultations. We would like to caution that both minimalistic and comprehensive teleconsultation platforms may not adequately address the needs and preferences of patients. For all stakeholders, compatibility with previous infrastructure emerged as essential requirement for various reasons. Doctors preferred compatible systems because they would already familiar with existing systems, while administrators valued the convenience of having access to previous reports. IT professionals also valued the ease of integration in order to minimize

Table 8 Compliance features across three levels of maturity

Minimal level	Comprehensive level	Smart assist level
Not many features and ultimately it is the responsibility of the hospital administrator to set internal guidelines and monitor usage to ensure policies are followed	Identification and authentication features	Compliance with clinical standards such as ICD10 or ICD11, SNOMED CT, and LOINC
	Capability to integrate with of national resident identity numbers, mobile/email-based identification methods, and two-factor authentication	Safe and secure data capture and exchange through integration with IoT devices, third-party solutions such as labs and pharmacies
	Provision to obtain consent for one-time platform use or explicit consent documentation before each virtual encounter	As part of consent management enable anonymous or de-identified data sharing for analytics and clinical research
	Ensure encrypted data transmission, data backup, audit trails, encrypted data backup	Consent management system should consider the privacy requirements of illiterate, dependent, or marginalized patients, considering their socio-cultural and socio-economic background, and obtain consent from patients for the use of their data by third parties. It should also provide access to patients' relatives or other members of their circle of care, clinicians, and healthcare providers for cross-consultation, with encounter-specific consents communicated in a manner that the user can understand and special consideration given to potentially stigmatized conditions
	Protect the privacy of both doctors and patients	Manage the data life cycle in accordance with national guidelines
	Capability to integrate with drug databases interoperability standards and requirements, set by the organization	Facilitate secure, encrypted data storage for medico-legal and clinical audit purposes

potential issues during the adoption process. In addition to these factors, for level three platforms, accreditations and the ability to interface with other existing systems are also important considerations for stakeholders [15]. It is important to note that the capabilities of comprehensive teleconsultation platforms are also informed by national and organization-level policies, data exchange standards implemented by healthcare organizations [17, 37, 52], and regional healthcare needs.

6 Limitations

The present study has two notable limitations that should be taken into consideration when interpreting the results. Firstly, only teleconsultation platforms that were available in India at the time of the study were included in the analysis. Secondly, with the exception of one organization, all of the organizations that were part of the alliance were based in India, and all of the professionals who served on the committee were from India as well. Although, it should be noted that several of the nominated experts had experience with healthcare systems in other countries, this geographic and cultural focus may still be a limiting factor in the generalizability of the findings.

7 Conclusion

This paper is motivated by the need to have tools to evaluate telemedicine platforms. We looked at a critical health IT component, i.e. Teleconsultation system. COVID-19 has increased both availability and acceptability of teleconsultation platforms across the globe including the low and middle-income countries, such as India. We reviewed previous work relevant to evaluation of teleconsultation platforms/solutions and learnt that stakeholders' engagement was mostly limited or absent resulting in evaluation studies with majority focus on cost and clinical efficacy of the platform. We noted that there is a need to break away from clinical service standard teleconsultation evaluation for an evaluation framework that accounts for situational factors and the complex dynamics of tele healthcare delivery ecosystems. We convened a diverse group of stakeholders to evaluate the feature offerings of existing teleconsultation platforms in India. Drawing upon their feedback, we have formulated an evaluation framework designed for utilization by hospital administrators, healthcare providers, platform developers, and other stakeholders faced with a multitude of teleconsultation platform choices. The framework presented in this paper is intended to be viewed as a step in an ongoing process, rather than a static entity. While we are presenting one version of the framework, it should be seen as a milestone along this continuous process, which can take various forms depending on the context in which it is being developed or applied. The recognition of the temporal nature of the framework relates to the shift among agile developers from specifying requirements in terms of features to focusing on "user stories" and "user journeys". This framework provides the opportunity to understand the process of maturing and hence the evolving maturity of the teleconsultation framework (and therefore this is an agile framework for evaluating maturity) in line with the recent call among information systems usability researchers to prioritize the ongoing temporal flow of IS phenomena ([8], p. 424).

Acknowledgements We wish to extend our acknowledgment to the reviewers team who played a crucial role in the development of manuscript. We express our thanks to the National Resource Centre for EHR Standards (NRCeS), the College of Healthcare Information Management Executives—India (CHIME-INDIA), the Consortium of Accredited Healthcare Organisations (CAHO), the Healthcare Information and Management Systems Society (India) (HIMSS-INDIA), and X-Scale Innovations. Their expertise and commitment have been instrumental in shaping our work.

References

1. Alkmim MB, Marcolino MS, Maia JX, Pessoa CG, Machado E, Sousa L (2015) Clinical quality control of a large-scale teleconsultation service. Stud Health Technol Inform 216(2015):988
2. Alkmim MBM, Marcolino MS, Figueira RM, Sousa L, Nunes MS, Cardoso CS, Ribeiro AL (2015) Factors associated with the use of a teleconsultation system in Brazilian primary care. Telemed e-Health 21(6):473–483
3. Almathami HKY, Win KT, Vlahu-Gjorgievska E (2022) Development and validation of a new tool to identify factors that influence users' motivation toward the use of teleconsultation systems: a modified Delphi study. Int J Med Inform 157(2022):104618
4. Andargoli AE, Scheepers H, Rajendran D, Sohal A (2017) Health information systems evaluation frameworks: a systematic review. Int J Med Inform 97(2017):195–209
5. Andersen RS, Vedsted P (2015) Juggling efficiency. An ethnographic study exploring healthcare seeking practices and institutional logics in Danish primary care settings. Soc Sci Med 128:239–245
6. Bakken S (2022) Centering the patient in informatics applications, pp 1027–1028
7. Barrett D (2017) Rethinking presence: a grounded theory of nurses and teleconsultation. J Clin Nurs 26(19–20):3088–3098
8. Baygi RM, Introna LD, Hultin L (2021) Everything flows: studying continuous socio-technological transformation in a fluid and dynamic digital world. MIS Q 45(1):423–452
9. Bergrath S, Czaplik M, Rossaint R, Hirsch F, Beckers SK, Valentin B, Wielpütz D, Schneiders M-T, Brokmann JC (2013) Implementation phase of a multicentre prehospital telemedicine system to support paramedics: feasibility and possible limitations. Scand J Trauma Resuscit Emerg Med 21(1):1–10
10. Berwick D, Fox DM (2016) "Evaluating the quality of medical care": Donabedian's classic article 50 years later. Milbank Q 94(2):237
11. Bestsennyy O, Gilbert G, Harris A, Rost J et al (2021) Telehealth: a quarter-trillion-dollar post-COVID-19 reality. McKinsey & Company, p 9
12. Bezerra GMF, de Lucena Feitosa SE, Catunda JGV, Graça CNS, de Aquino PL, Neto AGB, da Silva Junior GB (2022) Telemedicine application and assessment during the COVID-19 pandemic. Stud Health Technol Inform 290:854–857
13. Bhat KS, Jain M, Kumar N (2021) Infrastructuring telehealth in (in)formal patient-doctor contexts. Proc ACM Hum-Comput Interact 5(CSCW2):1–28
14. Bhatt SP, Patel SB, Anderson EM, Baugh D, Givens T, Schumann C, Sanders JG, Windham ST, Cutter GR, Dransfield MT (2019) Video telehealth pulmonary rehabilitation intervention in chronic obstructive pulmonary disease reduces 30-day readmissions. Am J Respir Crit Care Med 200(4):511–513

15. Birkmire-Peters DP, Whitaker LA, Peters LJ (1997) Usability testing for telemedicine systems: a methodology and case study. In: Proceedings of the human factors and ergonomics society annual meeting, vol 41. Sage, Los Angeles, CA, pp 792–796
16. Chandwani R, Kumar N (2018) Stitching infrastructures to facilitate telemedicine for low-resource environments. In: Proceedings of the 2018 CHI conference on human factors in computing systems, pp 1–12
17. Chib A, van Velthoven MH, Car J (2015) mHealth adoption in low-resource environments: a review of the use of mobile healthcare in developing countries. J Health Commun 20(1):4–34
18. Choemprayong S, Charoenlap C, Piromsopa K et al (2021) Exploring usability issues of a smartphone-based physician-to-physician teleconsultation app in an orthopedic clinic: mixed methods study. JMIR Hum Factors 8(4):e31130
19. Clarke M, Thiyagarajan CA (2008) A systematic review of technical evaluation in telemedicine systems. Telemed e-Health 14(2):170–183
20. Coiera E (2003) Guide to health informatics. CRC Press
21. Corbie-Smith G, Wynn M, Richmond A, Rennie S, Green M, Hoover SM, Watson-Hopper S, Nisbeth KS (2018) Stakeholder-driven, consensus development methods to design an ethical framework and guidelines for engaged research. PLoS ONE 13(6):e0199451
22. Currie LM (2005) Evaluation frameworks for nursing informatics. Int J Med Inform 74(11–12):908–916
23. Czekierda Ł, Malawski F, Wyszkowski P (2015) Holistic approach to design and implementation of a medical teleconsultation workspace. J Biomed Inform 57(2015):225–244
24. DeChant HK, Tohme WG, Mun SK, Hayes WS, Schulman KA (1996) Health systems evaluation of telemedicine: a staged approach. Telemed J 2(4):303–312
25. Di Cerbo A, Morales-Medina JC, Palmieri B, Iannitti T (2015) Narrative review of telemedicine consultation in medical practice. Patient Prefer Adher 9(2015):65
26. Dixon RF, Stahl JE (2008) Virtual visits in a general medicine practice: a pilot study. Telemed e-Health 14(6):525–530
27. Esterle L, Mathieu-Fritz A (2013) Teleconsultation in geriatrics: impact on professional practice. Int J Med Inform 82(8):684–695. https://doi.org/10.1016/j.ijmedinf.2013.04.006
28. Farzandipour M, Nabovati E, Tadayon H, Jabali MS (2021) Usability evaluation of a nursing information system by applying cognitive walkthrough method. Int J Med Inform 152:104459. https://doi.org/10.1016/j.ijmedinf.2021.104459
29. Ftouni R, AlJardali B, Hamdanieh M, Ftouni L, Salem N (2022) Challenges of telemedicine during the COVID-19 pandemic: a systematic review. BMC Med Inform Decis Mak 22(1):207. https://doi.org/10.1186/s12911-022-01952-0
30. Gackowski A, Czekierda Ł, Chrustowicz A, Cała J, Nowak M, Sadowski J, Podolec P, Pasowicz M, Zieliński K (2011) Development, implementation, and multicenter clinical validation of the TeleDICOM—advanced, interactive teleconsultation system. J Digit Imaging 24(3):541–551. https://doi.org/10.1007/s10278-010-9303-8
31. Gallerini S, Marsili L, Groccia V, Bartalucci M, Innocenti E, Marotti C, Pieri S, Plewnia K, Scarpini C, Keeling EG, Gregorio M, Geraci S, Zocchi M, Cirinei M, De Stefano T, Galassi S, Martini G, Tassi R, Bracco S, Cerase A, Dami S, Panzardi G, Breggia M, Marconi R (2020) Appropriateness, safety, and effectiveness of "drip and ship" teleconsultation model in Southeastern Tuscany: a feasibility study. Neurol Sci 41(10):2961–2965. https://doi.org/10.1007/s10072-020-04446-x
32. Georgsson M, Staggers N, Årsand E, Kushniruk A (2019) Employing a user-centered cognitive walkthrough to evaluate a mHealth diabetes self-management application: a case study and beginning method validation. J Biomed Inform 91(2019):103110
33. Glasgow RE, Vogt TM, Boles SM (1999) Evaluating the public health impact of health promotion interventions: the RE-AIM framework. Am J Public Health 89(9):1322–1327
34. Greenhalgh T, Vijayaraghavan S, Wherton J, Shaw S, Byrne E, Campbell-Richards D, Bhattacharya S, Hanson P, Ramoutar S, Gutteridge C et al (2016) Virtual online consultations: advantages and limitations (VOCAL) study. BMJ Open 6(1):e009388

35. Grenier Ouimet A, Wagner G, Raymond L, Pare G (2020) Investigating patients' intention to continue using teleconsultation to anticipate postcrisis momentum: survey study. J Med Internet Res 22(11):e22081
36. Idris H (2021) Utilization of teleconsultation: mitigation in handling mental disorders in the COVID-19 era. Int J Ment Health Addict 19(6):2320–2322
37. Intan Sabrina M, Defi IR (2021) Telemedicine guidelines in South East Asia—a scoping review. Front Neurol 11:581649
38. James D, Warren-Forward H (2015) Research methods for formal consensus development. Nur Res 22(3)
39. Kamal SA, Shafiq M, Kakria P (2020) Investigating acceptance of telemedicine services through an extended technology acceptance model (TAM). Technol Soc 60(2020):101212
40. Kea B, Sun BC-A (2015) Consensus development for healthcare professionals. Intern Emerg Med 10:373–383
41. Khandelwal R, Kolte A, Rossi M (2022) A study on entrepreneurial opportunities in digital health-care post-COVID-19 from the perspective of developing countries. Foresight 24(3/4):527–544
42. Kidholm K, Ekeland AG, Jensen LK, Rasmussen J, Pedersen CD, Bowes A, Flottorp SA, Bech M (2012) A model for assessment of telemedicine applications: mast. Int J Technol Assess Health Care 28(1):44–51
43. Li X, Rai A, Krishnan G (2020) Designing cost-effective telemedicine camps for underprivileged individuals in less developed countries: a decomposed affordance-effectivity framework. J Assoc Inf Syst 21(5):3
44. Light B, Burgess J, Duguay S (2018) The walkthrough method: an approach to the study of apps. New Media Soc 20(3):881–900
45. Lopez-Villegas A, Catalan-Matamoros D, Peiro S, Lappegard KT, Lopez-Liria R (2020) Cost-utility analysis of telemonitoring versus conventional hospital-based follow-up of patients with pacemakers. The NORDLAND randomized clinical trial. PLoS ONE 15:1–17. https://doi.org/10.1371/journal.pone.0226188
46. Lu W, Hou H, Ma R, Chen H, Zhang R, Cui F, Zhang Q, Gao Y, Wang X, Bu C et al (2021) Influencing factors of patient satisfaction in teleconsultation: a cross-sectional study. Technol Forecast Soc Change 168(2021):120775
47. Maarop N, Win KT (2012) Understanding the need of health care providers for teleconsultation and technological attributes in relation to the acceptance of teleconsultation in Malaysia: a mixed methods study. J Med Syst 36:2881–2892
48. Maeder A, Poultney N (2016) Evaluation strategies for telehealth implementations. In: 2016 IEEE international conference on healthcare informatics (ICHI). IEEE, pp 363–366
49. Mahatody T, Sagar M, Kolski C (2010) State of the art on the cognitive walkthrough method, its variants and evolutions. Int J Hum Comput Interact 26(8):741–785
50. Mahnke CB, Jordan CP, Bergvall E, Person DA, Pinsker JE (2011) The Pacific Asynchronous TeleHealth (PATH) system: review of 1,000 pediatric teleconsultations. Telemed e-Health 17(1):35–39
51. Otto L, Whitehouse D, Schlieter H (2021) Scaling up telemedicine initiatives: requirements for a new telemedicine maturity model. In: Research anthology on telemedicine efficacy, adoption, and impact on healthcare delivery. IGI Global, pp 240–255
52. Pai R (2021) Analysing the acceptability of a mHealth patient application for cross border data exchange by end-users
53. Park HS, Kim KI, Soh JY, Hyun YH, Lee BE, Lee JH, Jo JG, Lee HC, Kim HS (2020) Development and operation of a video teleconsultation system using integrated medical equipment gateway: a national project for workers in underserved areas. J Med Syst 44(11):194
54. Paul DL, McDaniel RR (2016) Influences on teleconsultation project utilization rates: the role of dominant logic. BMC Med Inform Decis Mak 16(1):1–18
55. Petcu R, Kimble C, Ologeanu-Taddei R, Bourdon I, Giraudeau N (2017) Assessing patient's perception of oral teleconsultation. Int J Technol Assess Health Care 33(2):147–154

56. Plumb JJ, Hains I, Parr MJ, Milliss D, Herkes R, Westbrook JI (2017) Technology meets tradition: the perceived impact of the introduction of information and communication technology on ward rounds in the intensive care unit. Int J Med Inform 105(2017):49–58
57. Rimner T, Blozik E, Begley C, Grandchamp C, von Overbeck J (2011) Patient adherence to recommendations after teleconsultation: survey of patients from a telemedicine centre in Switzerland. J Telemed Telecare 17(5):235–239
58. Rogers H, Madathil KC, Agnisarman S, Narasimha S, Ashok A, Nair A, Welch BM, McElligott JT (2017) A systematic review of the implementation challenges of telemedicine systems in ambulances. Telemed e-Health 23(9):707–717
59. Saechow S, Kamolphiwong S, Chandeeying V (2014) Web-based teleconsultation for clinical diagnosis. In: 2014 international conference on electronics, information and communications (ICEIC). IEEE, pp 1–2
60. Smith A, Ghosh K, Joshi A (2003) Usability and HCI in India: cultural and technological determinants. HCI Int (Crete)
61. Tchero H, Noubou L, Becsangele B, Mukisi-Mukaza M, Retali G-R, Rusch E (2017) Telemedicine in diabetic foot care: a systematic literature review of interventions and meta-analysis of controlled trials. Int J Lower Extrem Wounds 16(4):274–283
62. Toledo FG, Triola A, Ruppert K, Siminerio LM (2012) Telemedicine consultations: an alternative model to increase access to diabetes specialist care in underserved rural communities. JMIR Res Protoc 1(2):e2235
63. Torre Díez I, Alonso SG, Cruz EM, Franco MA (2019) Measuring QoE of a teleconsultation app in mental health using a pentagram model. J Med Syst 43(7):1–5
64. Uhl-Bien M (2021) Complexity and COVID-19: leadership and followership in a complex world. J Manage Stud 58(5):1400
65. Van Dyk L, Schutte CSL (2013) The telemedicine service maturity model: a framework for the measurement and improvement of telemedicine services. Telemedicine 2013:217–238
66. Viers BR, Lightner DJ, Rivera ME, Tollefson MK, Boorjian SA, Karnes RJ, Thompson RH, O'Neil DA, Hamilton RL, Gardner MR et al (2015) Efficiency, satisfaction, and costs for remote video visits following radical prostatectomy: a randomized controlled trial. Eur Urol 68(4):729–735
67. Yan Z, Guo X, Vogel DR (2016) Understanding dynamic collaboration in teleconsultation. Inf Technol Dev 22(1):152–167
68. Zahra SA (2021) International entrepreneurship in the post COVID world. J World Bus 56(1):101143

"So, Should I Walk Today or Not?" Understanding Concerns and Queries on Health and Fitness Among Indian Older Adults

Pallavi Rao Gadahad and Anirudha Joshi

Abstract This paper investigates concerns and queries of Indian older adults about their health and fitness (H&F), experts' views on the matter, gaps in existing technologies, and implications for designing effective technology-based solutions in this space. We conducted semi-structured interviews with 25 Indian older adults and consulted with five experts in the health and fitness domain. Our study revealed that older adults have several concerns and queries about their H&F and often lack clarity, holding misconceptions about maintaining a healthy lifestyle. The information they receive about H&F can sometimes be contradictory, inconsistent, out-of-context, or generic. Additionally, while older adults are motivated to maintain their health, they struggle to adhere to a routine. Personalized guidance and ongoing support are necessary to help older adults overcome these challenges. Our research identified many design opportunities under five key themes—personalization, localization, social interactions, information dissemination, accessibility and usability—that are crucial to consider when designing technology-based solutions for Indian older adults in this space. Finally, we developed a theoretical framework based on established adherence models and study findings, proposing strategies to promote adherence to H&F routines among Indian older adults.

Keywords Designing for Indian older adults · Technology in health and fitness · Adherence to health and fitness

P. R. Gadahad (✉) · A. Joshi
IDC School of Design, IIT Bombay, Mumbai, Maharashtra, India
e-mail: pallavi.rao@iitb.ac.in

A. Joshi
e-mail: anirudha@iitb.ac.in

A. Joshi and G. R. Sim (eds.), *Proceedings of the 14th Indian Conference on Human-Computer Interaction*, Studies in Computational Intelligence 1162,
https://doi.org/10.1007/978-981-97-4335-3_2

CCS Concepts: • **Human-centered computing → User studies; HCI theory, concepts and models**

1 Introduction

India is experiencing a growth in the percentage of older adults in its population. While increasing longevity is a matter to celebrate, studies have found multiple morbidities and disabilities linked to the advancement of age [53]. To address these concerns, the Government of India launched the National Programme for the Healthcare of the Elderly in 2021. This initiative aims to promote health education among older adults through various communication channels, emphasizing the importance of physical exercise, healthy habits, and stress reduction [12]. Despite the benefits of managing their Health and Fitness (H&F) through regular exercise, a healthy diet, and health check-ups, older adults face challenges in adhering to these activities.

On the other hand, older adults obtain a lot of information from numerous sources on matters related to H&F. Information on H&F is more accessible now than before with the increase of smartphones in the Indian consumer market. The number of smartphones in India grew from 34 million in 2010 to 931 million in 2022 [55]. Even though the percentage of Indian older adults using smartphones is much lesser (1.1% of overall users) than young adults [14], with the diffusion of information from various sources, there is a risk of getting contradictory or misleading information on H&F. This raises questions about whether Indian older adults are aware of what needs to be done to manage their H&F, what concerns or queries they have in this regard, what challenges they face, and what experts' views are on this matter. This paper aims to identify concerns and queries that Indian older adults might have about their H&F, experts' opinions on this matter, gaps in the existing health-related technologies, and opportunities for designing effective technology solutions. Specifically, we investigate the following research questions:

(1) What concerns and queries do Indian older adults have regarding their H&F?
(2) What suggestions do experts make in response to these concerns and queries, and what are their views on H&F of Indian older adults?
(3) How can technology solutions be designed to support H&F of Indian older adults?

To address the research questions, we conducted two sets of interviews. First, we interviewed 25 older adults to understand their concerns and queries regarding their H&F. Second, we interviewed five experts in the health and fitness domain to get their responses to these concerns and queries and to understand their perspective on the H&F of Indian older adults. Through this paper, we contribute in four ways. First, we highlight the concerns and queries Indian older adults have about their H&F. Second, we collate experts' responses to these concerns and queries, and their views on the H&F of Indian older adults. Third, based on the findings, we suggest ways to design technology solutions to support the H&F of Indian older adults. Finally, based on the study findings and existing adherence models, we propose a theoretical framework that outlines strategies to promote adherence to H&F routines among Indian older adults.

2 Background

2.1 *Aging and Adherence to H&F Routines*

Population aging is a global phenomenon, and India is no exception. The elderly population in India is projected to reach 194 million by 2031 from 138 million in 2021, representing a 41% increase over the decade, as reported by the National Statistical Office (NSO)'s Elderly in India 2021 report [11]. As a result, India, a relatively young country, is expected to become a rapidly aging society in the coming decades [53]. With increasing age, the risk of various disabilities and diseases, along with a decline in regular activities, also increases [23]. According to the Indian Council of Medical Research (ICMR) report [13], stroke is the leading cause of death and disability in India, with almost 700,000 deaths and 1.29 million cases reported in 2019. Most first-ever stroke cases were observed between 60 and 74 years of age. Physical inactivity, obesity, and an unhealthy diet are some factors causing these diseases.

Adherence, which refers to how people follow therapies intended to improve their health, has been extensively researched across various fields such as medicine, public health, nursing, pharmacology, and psychology [28]. This issue is particularly significant in the older adult population, where non-adherence can have severe consequences and result in higher healthcare costs [29]. Apart from medication adherence, studies have also emphasized the importance of adhering to specific diets [19, 24, 38] for older adults with health conditions such as diabetes and high blood pressure. Additionally, studies have demonstrated the significance of adhering to physical activities among older adults [43, 48, 49].

Although older adults require consistent adherence to H&F routines, many fail to comply with these activities regularly. For instance, irrespective of the specific guidelines by public health organizations like the World Health Organization (WHO) and the Indian Council of Medical Research (ICMR) on the recommended level of physical activities for older adults [47, 63], many older adults do not follow them [32, 54]. While we do not have detailed data on the physical activity patterns specifically of Indian older adults, a recent large-scale national survey conducted in 2020 [44] showed that 57% of the surveyed population (20–70 years old) do not meet the recommended physical activity regimen by WHO. This percentage was higher among people over 40 (63.4%) than those under 40 (51.7%).

2.2 Information on H&F

Indian older adults seem motivated to manage their H&F [45]. However, when it comes to knowing what needs to be done concerning diet, nutrition, and physical activities, they receive a lot of information from various sources besides their doctors. These could be media sources such as the Internet, television, printed materials, people from family and social circles, celebrity influencers, etc. Although these sources may create awareness of H&F, many sources frequently provide contradictory information. Moreover, most information is generic and not tailored to a person's age, health conditions, lifestyle, and other relevant factors.

It is worth noting that even among medical experts, disagreements can arise. For instance, a seemingly straightforward question such as *"Should people with diabetes eat more wheat and millet than rice?"* can elicit varying responses depending on the source. While some studies suggest that wheat and millets have lower Glycemic Index[1] than rice [40, 62] and are preferred staples for those with diabetes [40]; there is also a study that did not find any differences in the glycemic index of wheat and rice-based mixed meals [41]. These conflicting perspectives are particularly relevant in India, where traditional practices such as Ayurveda[2] are widely used alongside conventional Allopathic[3] medicine.

Information technologies have the potential to provide relevant information on H&F and support Indian older adults in managing their health. Many health-related technologies (such as mobile health and fitness apps, wearable trackers, and exergames) are designed to motivate and assist individuals in living an active and healthy lifestyle. Several health-related technologies have become well-established in the consumer sector. For instance, the Indian wearables market saw record growth in the first quarter of 2022, with shipments crossing 13.9 million units [15]. App stores also provide a variety of mobile health and fitness apps. However, not many older adults use these devices [26, 27, 56, 60], including older adults in India [45]. One of the main reasons these technologies have less impact on older adults than younger generations is that they have not been designed and developed based on the needs and preferences of the aging population [23]. The involvement of older adults in design processes is scarce, and their experience with a given technology is rarely considered relevant for evaluation [58]. Additionally, Harrington et al. [21] argue that the adoption of technologies by older adults depends not only on the usability of the design but also on older adults' preferences and prior experiences, which can be discovered by a better understanding of older adults themselves.

Research in the area of aging and technology has been gaining importance. Innovations in HCI research of health-related technologies can potentially improve older adults' access to healthcare resources and support long-term independence [21]. Past studies in this space have looked into how health-related technologies help in

[1] Glycemic index is a number which gives an idea about how fast our body converts the carbohydrates in a food into glucose [6].

[2] Traditional medicine system with historical roots in the Indian subcontinent [3].

[3] A term used to define science-based modern medicine [2].

the health monitoring of older adults (e.g., [57]) and enable behavior change (e.g., [54]). Some studies have identified the enablers [30, 34, 35] and barriers [56] that older adults face in using health-related technologies. Additionally, some studies have highlighted that the existing health-related technologies do not comply with the needs of the aging population (e.g., [23, 52]). Few studies from Europe and the USA (e.g., [22, 61]) have given recommendations for designing specific health-related technologies after understanding the needs and preferences of older adults.

While "healthy aging" is universally relevant, cultural differences in lifestyle, traditions, and customs mean that older adults' needs, preferences, concerns, challenges, and healthcare practices vary across cultures. Although there has been considerable research on health-related technologies for managing routine H&F activities among older adults, much of it has focused on older adults from China [56] and other developed countries (e.g., [22, 30, 34, 35, 61]). There is a dearth of research on older adults in developing countries, specifically in the context of Indian culture and lifestyle.

3 Interviews with Older Adults

3.1 Method

We conducted semi-structured interviews with 25 older adults aged 60 and above (with a mean age of 70.80 years and a standard deviation of 6.6). The interviews were conducted by the first author with experience in conducting qualitative interviews. The participants were selected through snowball sampling, whereby we initially reached out to a group of older adults within our social network and then asked them to connect us with other potential participants. The selected participants were older adults from middle-class families from urban or metropolitan cities.[4] Broadly, we define middle-class families based on (a) residential locations—people residing in apartments and housing complexes and (b) lifestyle indicators—people with access to education, healthcare facilities, and stable livelihoods.

As we would want the participants to discuss their physical activities, we did not select older adults who are bedridden, advised by doctors not to exercise, or are under supervised exercise by physiotherapists. This study was approved by the Ethics Committee of our university.

To collect data from our participants, we used a combination of face-to-face and remote interviews conducted over the phone. Prior to the interviews, we provided each participant with a brief overview of the study's objectives and ensured them that their identities would remain anonymous. Once we had obtained their consent, we followed our interview schedule, which covered topics such as their current H&F activities, level of awareness about H&F, and sources of information on H&F. In addition, we asked participants to imagine themselves in front of a panel of experts, which

[4] Classification of cities based on population; Urban population—100,000–999,999; Metropolitan population—1,000,000 and above [4].

included physiotherapists, geriatricians, doctors, dietitians, and fitness trainers, and to share their concerns and queries related to their H&F based on their personal experiences. By asking participants to envision themselves in front of a panel of experts, we aimed to create a context closely resembling real-life situations where individuals seek medical advice. Also, by framing the exercise as an interaction with experts rather than researchers, we sought to minimize the bias as participants may feel more comfortable expressing genuine concerns they might hesitate to disclose otherwise.

Based on the participant's preference, the interviews were conducted in either Kannada, Hindi, or English and lasted approximately 30–45 min. We did not provide any incentives to the participants.

3.2 Data Analysis

The data was coded by both authors, with experience in qualitative data analysis. We transcribed and coded each interview individually, following the concept-driven coding procedure, also known as deductive coding, as recommended by Gibbs [20]. Gibbs [20] explains that in this coding approach, the categories (or concepts) the codes represent may come from research literature, previous studies, topics in the interview schedule and so on. In our study, we analyzed the data by coding it into categories that were derived from the topics covered in our interview schedule. Based on our analysis, we present the findings in three main categories: **Awareness about H&F**, **Information sources on H&F**, and **Concerns and queries about H&F**.

3.3 Findings

Our study consisted of 25 participants, including 13 males and 12 females. Four of the participants identified as homemakers, while the remaining 21 had previously worked in various roles and were currently retired. Of the participants, 14 fell within the age range of 60–70, 8 were between 70 and 80, and 3 were above 80. The participants resided in urban areas (12) or metropolitan cities (13). All participants lived with either their spouse or alone if their spouse had passed away, and none lived with their children or other relatives.

3.3.1 Awareness About H&F

All the participants knew the importance of managing their H&F to lead an active and healthy life. This awareness appeared to be influenced by various factors, such as their children's insistence, influence from their social circle, media, or celebrities, and advice from their family doctors. The primary motivation for most participants was to remain independent and self-sufficient. As many participants mentioned, they *"do not want to rely on others"*. For a few participants whose children resided abroad, maintaining good H&F was crucial because they did not want to *"burden their children"* in any way.

When asked about their eating habits, some participants said they try to eat healthy food and eat in moderation. One of them specifically said, *"less salt, less sugar, and no junk food"*. However, many of them seem confused about what constitutes a healthy diet. For instance, one participant said that she gets *"a lot of healthy recipes through her WhatsApp groups, but most of the ingredients mentioned are unavailable in the vicinity"*.

Although most participants stated that they regularly went for walks, further probing revealed that they often missed their physical activity routine due to various reasons such as rainy seasons, ailments, guests at home, and friends not showing up. According to the Indian national guidelines for fitness [47], older adults should engage in at least 150 min of moderate-intensity physical activity per week or at least 75 min of vigorous-intensity physical activity per week or an equivalent combination of both. For additional health benefits, the guidelines suggest increasing moderate-intensity physical activity to 300 min per week, engaging in 150 min of vigorous-intensity aerobic physical activity per week, or combining both. The guidelines also suggest muscle-strengthening and aerobic activities. However, participants were neither aware of these guidelines nor did they track their physical activities.

Going for regular health check-ups was not very common among the participants. Before COVID-19, some were going for regular health check-ups. However, after COVID-19, they have yet to return to that routine. Some participants rely on dietary supplements such as Calcium and Vitamins to address their health needs, typically consuming them when experiencing fatigue or muscle pain. For instance, one participant said *"Whenever I feel very tired or have pain in my legs, I start a course of Vitamin D"*. Additionally, while Allopathy is the preferred method of treatment for most health concerns, some participants turn to Ayurveda due to its perceived lack of adverse effects.

Participants know how crucial it is to manage their H&F, and they have some understanding (obtained from various sources) about it. However, many find adhering to a routine (in physical activities or diet) challenging. For instance, many participants said it was *"difficult to go for regular walks during family get-togethers"*. Also, many do not monitor their food intake *"during festivals"*.

Participants do not fully comprehend what must be done to manage their H&F. For example, one of them said, *"maintaining a routine in food, exercise, sleep is the only way to lead a healthy and active life"*. However, he was unaware of how much he needed to exercise, how much sleep would be sufficient for him and what quantities of food he needed to consume daily. He said he is happy *"as long as he is healthy"*. Most participants echoed similar opinions. We describe the challenges faced by the participants in managing their H&F, their concerns and queries in the "concerns and queries" section below.

3.3.2 Information Sources on Health and Fitness

Participants follow their family doctor's advice on H&F. However, they do not approach them for every concern or query related to H&F. While highlighting the limited time doctors give to their patients, one participant said, *"My doctor's clinic is always crowded, and he does not have time to answer all my concerns and queries. It is understandable because there is a long queue of people waiting for him."*.

All participants in our study used a smartphone, but none used any health and fitness app or a fitness tracker. Those aged 60–70 know how to search YouTube videos when they have concerns about their H&F. Some search for videos of well-known physicians in the area. Participants also obtain health and fitness-related information through their social circle, primarily via WhatsApp forwards, and television programs. A few participants expressed a lack of trust in the information provided by technology sources and preferred books and printed materials. However, most participants felt that their queries remained unanswered regardless of the source of information.

3.3.3 Concerns and Queries About Health and Fitness

Participants discussed their concerns and asked queries about their H&F. They highlighted the inconsistencies and contradictions in the information they get and raised concerns about the difficulties in choosing the correct information. After collecting all the concerns and queries from the participants, we removed duplicate ones and combined some similar ones. Participants' concerns and queries were on their diet and nutrition, physical activities, and general health. Analyzing their concerns and queries, we identified three categories: **Know-What**, **Know-Why**, and **Know-How**. These categories indicate participants' lack of knowledge in finding factual information, understanding underlying reasons, and following procedures, respectively, about their H&F.

Know-What: Concerns and queries in this category involve finding factual information (**Descriptive/Declarative knowledge**). Most queries fall under this category. For instance, *"Are eggs good for health, given our age? Does it increase cholesterol levels?"*, *"Is the BMI index the same for Indians as in the West? Alternatively, is it different for different cultures?"*. Further, there were queries about consuming *ghee*[5]*—"Is ghee bad for health? Ayurveda practices propagate ghee, but as per Allopathy, ghee raises cholesterol levels."*.

Some participants have borderline diabetes or a family history of diabetes. Their queries were on which food to consume to avoid diabetes or to keep it in control. Some of these queries were, *"Which fruits should diabetic people avoid? Are all fruits*

[5] Ghee is a clarified butter. It is commonly used in Indian and Middle Eastern cuisine, as a traditional medicine, and for religious rituals [5].

bad for diabetic people?", *"Is rice bad for people with diabetes? Are millets a better option for people who have diabetes? Are all carbohydrates bad for diabetes?"*.

Some of the queries related to physical activities were—*"Should I work out every day?"*, *"What exercises are safe, given my age?"*, *"Is it better to walk on a hard surface (such as a road) or a soft surface (such as on a ground)?"*. Further, some participants were unsure about exercising while having some pain or minor ailments. Their queries were, *"Is it ok to walk/jog/run when I have leg pain? Or should it be stopped completely?"*, *"Is it ok to do yoga or other exercises with a mild ailment like a backache? What type of yoga can be done when I have back pain/other ailments?"*. Some participants were interested in learning new exercises but were apprehensive about starting independently. Their queries were like *"I want to start a new exercise/practice yoga. Am I too old? Is there any age to start with a new exercise?"*

Many participants had queries on following Ayurvedic herbs and medications. For instance, some participants had started taking *Ashwagandha*[6] and *Guduchi*[7] after COVID-19 to boost immunity. There are studies (e.g., [16, 51]) showing that these herbs are effective in boosting immunity against COVID-19. One of the participants asked whether *"it is ok to take these herbs regularly?"*. Another participant, who has a long-term urinary tract infection, is taking Allopathic medications. Recently, she began using Ayurvedic medications as well, as per her friend's advice. She has kept her use of Ayurvedic medications a secret from her Allopathic doctor. She is unsure if it is ok to take both medications at once and asked, *"Is it ok to mix Ayurvedic and Allopathic medicine?"* There were also questions on drinking *kadha*[8], *"Does drinking kadha have any health benefits?"*

There were queries on health check-ups. A few participants went for health check-ups prior to COVID lockdowns. However, they did not know *"How frequently should one do health check-ups?"*. Also, there were concerns about health packages offered at the hospitals. They were confused about which package to select and asked, *"What tests should be done during regular health check-ups?"*

Several participants had heard mixed opinions about taking dietary supplements (such as Calcium, Vitamin-D, and Vitamin B Complex). Their queries were, *"What supplements do we need to take? Is it ok to take supplements without consulting doctors?"*, and *"Must I take Vitamin D supplements, or is it enough to walk under the sun?"*

Know-Why: In many cases, participants not only wanted to know what they should do, but also why they should do those things. While not too many, some concerns and queries in this category were about knowing the underlying reason or principle (**Conceptual knowledge**). Some of them were, *"Why should I keep a target of 10K steps every day?"*, *"Why is Intermittent fasting considered beneficial? Is it suitable for us, given our age?"*, *"I do not have any ailments as such. However, I do not*

---

[6] Medical herb in Ayurveda with a scientific name *Withania somnifera* [10].

[7] Medical herb in Ayurveda with a scientific name *Tinospora cordifolia* [9].

[8] Ayurvedic decoction made from a mixture of spices and herbs [36].

have much appetite. Why don't I feel hungry even after walking?", "Why do I get significantly less sleep (only about 4 hrs each day)?"

Know-How: Concerns and queries in this category involved asking for instructions or a procedure to follow (**Procedural knowledge**). For instance, *"How do I plan my diet?", "How do I know which food (that is available in my vicinity) can be categorized as nutritious? From where will I get the correct information, apart from doctors?", "Given my age and health condition, how long should I walk/work out/exercise? How should I plan my physical activities?", "Have heard that strength training is good even for older adults. How should I do it?", "How can I improve my sleep schedule?"*. In some cases, participants were already knowing or doing something, but they were not sure if they were following the procedure correctly. For example, *"I have learnt pranayama on my own. I am not sure whether I am doing it correctly. How will I know the correct procedure?"*.[9]

4 Interviews with Experts

4.1 Method

To answer the concerns and queries of our older adult participants, we consulted five experts:

(1) Dietitian and clinical nutritionist (PhD in Home Science—Food and Nutrition, Female, 55 years)
(2) Physiotherapist (Bachelor of Physiotherapy, Female, 38 years)
(3) Physician (MD in Pharmacology, Male, 59 years)
(4) Geriatrician (MD in General Medicine and Public Health, Female, 70 years)
(5) Occupational therapist (PhD in Geriatrics, Female, 45 years).

We carefully selected these five experts based on their expertise in the areas of diet and nutrition, physical activities, general health, and geriatric care. Among them were two practitioners (the dietitian and the physiotherapist), while the other three were professors from various medical colleges, all with several years of experience in their respective fields. The first author conducted semi-structured interviews with these experts after briefing our study's objective and obtaining their consent. We had taken approval from our university's ethics committee to interview these experts.

During the interviews, we directed older adult participants' queries to the relevant expert—the dietitian for diet and nutrition, the physiotherapist for physical activities, and the physician for general health concerns. We discussed all three topics with the geriatrician. In addition, based on the findings from the interviews with older adults, we asked each expert about frequent misconceptions about H&F among older adults, common points of disagreement among experts, the need for personalization

[9] Yogic practice of focusing on breath [8].

of advice, and the role of technology in supporting H&F of older adults. Further, we interviewed the occupational therapist about general wellness programs for older adults, how these programs are conducted, the effectiveness of these programs, and the role of technology in conducting these wellness programs. All the interviews were conducted remotely over a phone call. The interviews were conducted in English. Each interview took one to two hours.

4.2 Data Analysis

Each interview was transcribed and coded individually by both the authors. Similar to the data analysis of the previous interviews, we used a concept-driven coding procedure [20] to analyze these interviews. Gibbs [20] highlights that along with an initial list of codes, researchers might need to *"amend the list of codes during analysis as new ideas and new ways of categorizing might be detected in the data"*. Accordingly, from the interview data, we not only derived the categories from the topics in the interview schedule but also discussed and agreed on the new categories. The six categories derived from the data are: **Answers to older adults' concerns and queries**, **Disagreements among experts**, **Misconceptions among older adults**, **Research in local medical practices**, **Personalized guidance and support**, and **Socialization and community involvement**. In addition to this, based on the findings from both sets of interviews, we have identified several design opportunities. Below we organize the findings from interviews with experts around the six categories and intersperse the design opportunities where they are relevant.

4.3 Findings

4.3.1 Answers to Older Adults' Concerns and Queries

The experts provided in-depth responses to the queries, offering detailed information on each issue. For example, when addressing questions related to diet and nutrition, the geriatrician and dietitian elaborated on how older adults can achieve a well-rounded diet by incorporating foods from the five essential food groups: vegetables, fruits, whole grains, protein, and dairy. The conversations also delved into specific details about each food group, including serving sizes, recommended consumption, and the importance of each group for this demographic. In response to questions about physical activities, the physiotherapist recommended that healthy older adults aim for between 6000 and 10,000 steps per day. The geriatrician stressed that *"just walking for 30 minutes is not enough"* and asserted that older adults need to remain active all day. She said, *"it is important to do other kinds of exercises such as resistance training, yoga, and meditation, twice or thrice a week"*.

Although the experts answered most of the queries, some remained unanswered due to insufficient evidence. For instance, when asked about the efficacy of Ayurvedic herbs, the physician (who practices Allopathic medicine) stated that it was challenging to answer such questions without sufficient evidence.

In addition, the experts emphasized that responses to individual queries would differ depending on each older adult's context and health status. For example, when addressing questions about walking with leg pain, the physiotherapist noted that it is crucial to identify the type of pain experienced. She stated, *"If it is muscle pain, then it is generally safe to walk, but any other nagging pain needs to be diagnosed"*. The experts expressed similar sentiments while answering some of the queries such as *"whether to walk on hard or soft surfaces?"*, *"how many hours of sleep is required?"*, *"which supplements to take?"*, etc. Experts felt that answers to many such queries vary significantly between older adults.

Design Opportunities: Several of our older adult participants turned to YouTube to seek answers to their H&F-related questions. However, they found that the information available on the platform was often generic and not tailored to their specific needs. Although there are several mobile health and fitness apps and wearable trackers available in the market, they tend to focus primarily on monitoring activities such as sleep, calorie intake, and steps walked and do not offer much in terms of personalized, actionable advice. Design opportunities in this space could involve utilizing emerging technologies like chatbots or virtual assistants to provide H&F-related information and advice in a conversational manner. Further, recent research has explored the use of Artificial Intelligence (AI) enabled mobile chatbots to provide personalized responses to users seeking health and fitness advice [39, 50]. More research is needed to develop personalized conversational agents, capable of answering different types of questions, including Know-What, Know-Why, and Know-How queries, similar to the questions posed by our participants. Understanding the types of questions that an agent is expected to handle can help inform the design and development of the agent.

4.3.2 Disagreements Among Experts

Some disagreements among experts were evident from our interviews, either in the form of contradictory recommendations or giving the same recommendation, but for a totally different reason. While the dietitian said that it is *"OK"* for older adults with diabetes to eat fruits, the physician cautioned that older adults with diabetes should control their fruit intake. The dietitian considered that *"fasting is not advisable for older adults"* because she feels fasting, is *"abusing the body by minimizing nutrients"*. On the other hand, the physician cannot suggest fasting to his patients because *"it is not backed up by any theory or proven in medical sciences"*. He said these practices are left to the patients because there is no scientific basis. Even though both these experts agree on not recommending fasting, they do not seem to have the exact reason for their agreement.

Experts also highlighted the known disagreements in the medical community. For instance, the dietitian mentioned that, generally, doctors do not think of diet and nutrition in their treatment. She said, *"doctors focus only on the disease, ignoring the holistic picture which involves diet and nutrition, and this attitude causes problems"*. Even the geriatrician said that *"doctors cannot tell what to eat as they are not qualified in diet and nutrition, and nutritionists have no medical background; thus, there is a gap between these two professions"*. The physiotherapist echoed similar thoughts. She raised concerns about how some orthopedics do not recommend physiotherapy for older adults, even when she thought it necessary.

In addition, the physician explained that sometimes experts may have diverging opinions or interpretations of the same situation, despite standardized protocols for treatment. This variability can stem from differences in past experiences and approaches among healthcare professionals. The geriatrician also pointed out that evidence-based cases often require at least five years to become established within the medical community. Consequently, doctors may rely on their own experiences in treating such cases, which can contribute to disagreements among medical experts.

Design Opportunities: In India, disagreements among experts are prevalent due to diverse medical practices. While some H&F information may have scientific validity, others may have been passed down through generations as part of local traditions. However, their benefits may not have been demonstrated by contemporary science. This is particularly evident in developing countries with rich historical backgrounds. This creates a situation where an individual may encounter contradictory information and may not be certain about what they ought to do. Existing health-related "apps" sidestep these issues and completely disregard local practices. There is a design opportunity to contextualize local practices while incorporating scientifically verified information whenever possible. It is essential to acknowledge scientific disagreements when they exist while providing actionable insights to the user. This approach aims to empower individuals to make informed decisions that align with their cultural contexts, preferences, and beliefs and encourage the adoption of new behaviors where necessary. The information can be conveyed in various regional languages. Additionally, visuals such as diagrams, infographics, culturally relevant visual aids, and videos can be used to explain complex concepts. Establishing trust with older adults and reinforcing the credibility of the information provided can be done by citing authoritative sources, such as doctors and trusted agencies.

4.3.3 Misconceptions Among Older Adults

The experts commented that many older adults harbor misconceptions about H&F. The geriatrician shared her experience about how social media are responsible for spreading misconceptions. One of her colleagues gave a talk about cholesterol and lifestyle changes. In the talk, the doctor discussed the importance of regular exercise, eating in moderation, and controlling obesity. Further, the doctor also mentioned that if one does these activities regularly, one may not need to take any medication for cholesterol. Later, a short snippet of his recorded video was circulated, in which only a portion of the talk where the doctor says *"there is no need to take medications to control cholesterol"* was shown. This snippet was taken out of context. By watching this snippet, many older adults misinterpreted that no medication is required for cholesterol.

The physician said that as some people age, their misconceptions increase. The dietitian said, *"people just Google and start self-medication without understanding anything"* and raised concerns about older adults getting misinformation on nutrition from various technology platforms that might lead to misconceptions.

Design Opportunities: Our interviews show that older adults frequently receive contradictory, inconsistent, out-of-context, or generic information on H&F, leading to potential misconceptions. While this is a problem for all age groups, it may be particularly challenging for older adults. This sector encompasses a broad ecosystem of factors, from medical expertise to cultural influences, and involves various stakeholders. These complexities present significant challenges that extend beyond design. While we cannot solve these issues through technology design, we can offer older adults accessible and understandable information that they can use to navigate these complexities, empowering them to navigate this landscape, even when confronted with conflicting information. Furthermore, we can aim to establish a holistic framework that accounts for knowledge gaps in older adults, covering what they know, why they should know, and how they can apply this knowledge effectively. Therefore, it is important to communicate to older adults what to do (facts), how to do it (procedures), and how and why they work (concepts) aspects of H&F practices. Additionally, providing real-time progress can aid older adults in determining the effectiveness of their chosen approach. Existing technologies can be used to build these platforms, and the information design can incorporate the elements (e.g., local languages, visual aids) explained in the previous section.

4.3.4 Research in Local Medical Practices

The physician and the geriatrician highlighted that many of the medical protocols in India are based on global literature, and relatively fewer are based on local research. The physician cited an example of a community in Karnataka[10] with a particular enzyme deficiency. Though such cases are few, doctors need to know this while treating patients. However, there are no established theories behind it. Doctors learn about such issues through experience or their networks. He said, *"Indians are genetically different, but we do not have robust research to follow"*. The geriatrician said India's Allopathy research is localized to some extent and that there is ongoing research for ailments such as diabetes and cancer.

Speaking of localizing H&F, the geriatrician said that in India, Ayurvedic practices have many tips for healthy aging as it focuses on the holistic concept of wellness,

[10] A state in the southwest part of India.

which includes diet, nutrients, yoga, and mental health. However, she warned that rapid methods are sometimes adopted in Ayurvedic practices without much research. Also, many Ayurvedic practitioners do not record the cases. The physician said that research in the medical community takes time as many parameters (such as quantity, drug interactions, different groups of people, etc.) are considered in any medical experiment. Moreover, he says research in Ayurvedic practices does not study these details in-depth. He gave an example of an Indian herb ("*mulethi*[11]") which is considered good for immunity and cold in Ayurveda. However, there are not many studies conducted on different groups of people. Moreover, he emphasizes that, without proper prior research, trying any new herb or medication on people with ailments is not advisable.

Design Opportunities: Although many existing health-related technologies include standard localization features such as location and language translation, there is an opportunity to enhance these features by providing more contextual information. For example, in addition to identifying a user's location, the technology could also provide information on nearby fitness facilities (e.g., parks, trails, gyms) or events (e.g., walking groups, exercise classes). A database of local food and cultural practices can be built, and this information can be incorporated into personalized recommendations on H&F. Additionally, information on H&F based on local medical research and practices can also be provided. This kind of localization does not exist in the current H&F technologies.

4.3.5 Personalized Guidance and Support

One point echoed by all the experts is that there cannot be a one-size-fits-all approach to H&F for older adults. They stressed the importance of understanding the context and needs of every older adult and giving personalized treatments. The dietitian said that the diet she prescribes to her patients is personalized to their needs. The physiotherapist also noted that each older adult has a different medical condition and problem. More importantly, *"balance, flexibility, and neurological issues vary"* among older adults. Hence, physical activities for older adults need to be personalized. The occupational therapist discussed different activity thresholds for older adults and the need for personalizing physical activities. Further, the geriatrician said that whether it is about diet and nutrition or physical activities, *"everything depends on older adults' metabolism, weight, BMI and health conditions"*. Stressing that H&F matters are not the same for all older adults, she highlighted that in Ayurvedic practices, the focus is on treating individuals based on their *"Prakriti"*, which essentially is about personalizing wellness. However, the downside is that these practices are less researched and documented.

According to the dietitian, technology-based solutions offer general information and do not cater to personalizing diet and nutrition. She expressed her dissatisfaction

---

[11] Medical herb in Ayurveda with the scientific name *Liquorice/Glycyrrhiza glabra* [7].

with the current diet and nutrition apps, stating that most of these apps rely on calorie counting based on food types. However, calorie counts can vary for the same dish prepared by different people. While she acknowledged the usefulness of such apps in tasks like tracking physical activity, she believes they do not substantially support an individual's overall health and fitness goals.

Additionally, the occupational therapist who conducts wellness programs commented on how older adults are reluctant to use technology for physical activities. During her programs, she has noticed that many older adults prefer engaging in physical activity with a human being rather than relying on technology. She stated that older adults require more personal support than what a standalone app can offer, particularly during physical activity, as falls are a significant risk in this age group. The geriatrician emphasized the importance of older adults engaging in various exercises, such as aerobics, yoga, and dancing, to maintain their fitness. However, she has observed that older adults are hesitant to participate in new physical activities independently and require ongoing support, such as a human being talking to them about their H&F.

Design Opportunities: While digital technologies are extensively localized and personalized today, such localization and personalization are not yet common in health-related technologies. Localization and personalization came across as a huge opportunity from our findings, especially in the context of older adults. Achieving effective localization and personalization is neither straightforward nor a panacea. However, localization and personalization are essential because information (while it remains complex) needs to be user-actionable. We can draw inspiration from successful implementations of crowd-based, extensive localization (e.g., Google Maps) and search personalization (e.g., YouTube). Thus, while current technologies have achieved high personalization and localization, including in complex contexts of developing countries and emergent users, there are not many examples in the H&F domain, such as diet and exercise.

Experts went into considerable detail on each H&F issue, focusing on individual needs. There is much information on H&F available on digital platforms, including YouTube and fitness apps, which some of our participants used. Even if this information is helpful, it is generic. It is not easy to filter out what is relevant to the individual. It is even less useful if a user has a specific query. We believe there is an opportunity to provide granularly detailed information on health and fitness personalized to factors such as medical conditions, treatments, metabolism, weight, BMI, and food preferences. While some fitness apps already consider factors such as age, height, weight, physical activities, diet, and sleep (e.g., [1]), it is essential to incorporate health parameters, local context, and experts' guidance, especially for older adults, when designing diet and workout plans. This personalized approach could be achieved by developing new technologies that provide tailored feedback and recommendations based on the individual's health conditions, preferences, and goals.

Personalized support can be enhanced by incorporating human interventions into technologies. One example is using technology solutions to connect older adults with trainers who can monitor their progress and provide personalized interventions. Additionally, virtual coaching and support features can be integrated into fitness technologies to provide personalized guidance and feedback on exercise routines. These features could connect users with virtual coaches who can offer advice on exercising safely and preventing falls. Technologies can also be designed to detect older adults' gait and posture to provide real-time feedback on their balance and stability, thus reducing the risk of falls.

Personalized support can also be provided by various design features. For example, in a study on identifying and removing the barriers older adults face while using wearable activity trackers [46], researchers have used personalized "how-to" videos as a support feature to help participants use the trackers without any hassle. Similarly, many personalized support features, such as nudges (to remind), gamification elements (to engage and motivate), videos and walkthroughs (to assist using technologies), chatbots (to answer queries), etc., could be incorporated.

4.3.6 Socialization and Community Involvement

The geriatrician and occupational therapist emphasized that older adults prefer socialization during physical activities to stay active. According to the geriatrician, older adults need continuous external motivation and are happy to exercise in a social group. She said that as people age, their ability to do things independently reduces. Social groups in the form of communities motivate them to be physically active. The occupational therapist said her team conducts wellness programs that train many older adults in physical activities, mental health, and safety (especially avoiding falls). These programs are conducted by having tie-ups with associations, community centers, etc. She said participants enjoy these activities and regularly adhere to a schedule when performing in a group.

To promote healthy aging across India, the geriatrician suggested establishing community wellness platforms for older adults. She shared her experience running an offline platform for older adults in her city, where she and other medical professionals train and guide participants on exercises, healthy eating, and other aspects of healthy living. Although this group is small and limited to the city, she considers it successful in helping older adults adhere to their H&F activities.

Design Opportunities: Socialization has been shown to improve physical activity in older adults [59]. In a study on understanding the long-term usage of activity trackers, Kononova et al. [30] showed that increased socialization via activity trackers is an essential factor for long-term usage among older adults from the USA. Aligning with this, a study by Talukdar et al. [56] highlighted that "social influence" is a significant predictor in accepting wearable health technologies among older adults from China. Despite some fitness apps offering the ability to create or join groups, many still focus primarily on individual goals. To further support socialization and engagement among older adults, fitness apps could incorporate additional design features such as social prompting, tracking group member activities over a period of time, and gamification of group challenges.

A technology-enabled wellness platform could connect many stakeholders, including older adults, the community, elderly care centers, labs, pharmacies, clinics, hospitals, and experts from different domains (physiotherapists, dietitians, physicians, occupational therapists, geriatricians, etc.). The platform can motivate, conduct, and train older adults on their H&F. Creating such platforms may motivate older adults to learn about H&F activities from trusted peers who have successfully adopted them. The platform can enable experts to suggest interventions to individuals or groups. It can also enable the experts to track individuals' and groups' H&F activities over a longer term. Older adults can ask questions and get their doubts clarified. Experts can also learn from each other. The platform could incorporate gamification elements, rewards, and challenges to promote sustained engagement. As many people would be connected through this platform, it can run at a cost lower than an individual may otherwise need to pay to access these multiple services and consultations.

5 Discussion

This study is part of ongoing research about designing technology-based solutions to support adherence to H&F routines among middle-class older adults in India. When considering designing technology solutions to help this population adhere to a H&F routine in the Indian socio-cultural scenario, it is important to acknowledge the unique cultural values and practices that shape their daily lives. Even though many middle-class older adults now live independently due to changes in social structures, such as the shift away from extended families and relocation to metropolitan areas or other cities/countries, family ties are highly valued in this particular subculture. Traditionally, adult children are expected to care for their elderly parents. This cultural norm can significantly impact the design of technology solutions aimed at improving the H&F routines of this age group, as the involvement of family members and caregivers may be a critical factor in the success of such interventions.

In recent years, traditional Indian practices such as yoga and Ayurveda have gained popularity among educated Indians, owing to their potential health benefits. Therefore, it is important to consider how technology solutions can be integrated with these practices. Another important consideration when designing technology solutions for older adults in India is the digital divide. While Internet access is not a problem for this subculture (as they belong to middle-class families and have access to resources), technology literacy remains a concern. As a result, technology solutions designed for this population should prioritize accessibility and usability.

Our study findings showed that Indian older adults are motivated to maintain their H&F but may lack knowledge about managing it effectively. Additionally, adherence to H&F routines is often a significant challenge for this population. Therefore, technology solutions should empower older adults by providing them with the necessary tools and knowledge to take a more active and leading role in managing their H&F. Given this, there are significant implications for the design and research of technology-based solutions for this population, as explained below.

5.1 *Implications for Design*

We grouped the design opportunities into five key themes. These five themes are crucial in designing technology-based solutions for Indian older adults in this space.

1. **Personalization**

   (1) Personalization based on language, food preferences, health parameters
   (2) Personalization of guidance, feedback, nudges, recommendations and reminders on H&F activities.

2. **Localization**

   (a) Localization, based on location, local food habits, local traditions, local medical practices.

3. **Social Interactions**

   (a) Social features in health-related technologies—social prompting, tracking group activities, gamification of group challenges
   (b) Community platforms connecting older adults and domain experts.

4. **Information Dissemination**

   (a) Visualization of data collected on H&F activities—personalized dashboard, progress tracking, comparative analysis, goal visualization, interactive heatmaps
   (b) Communicating what-to-do, how-to-do, how-and-why they work on various H&F practices
   (c) Incorporating trusted agencies to show credibility of H&F information.

5. **Accessibility and Usability**

   (a) Incorporating tech support features—Q&A chatbots, how-to-videos, walkthroughs
   (b) Usable interactions such as voice interactions
   (c) Use of visuals such as diagrams, infographics, culturally relevant visual aids, videos.

While the five key themes may seem broad at a high level, the design opportunities discussed within each theme hold substantial relevance and contextual significance.

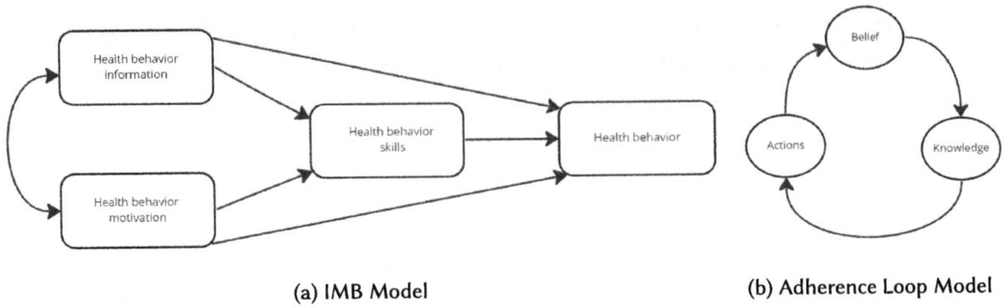

Fig. 1 IMB model [18] and adherence loop model [28]

5.2 Implications for Research

Adherence, as defined by Klein et al. [28] is *"taking a medication or performing a therapy as directed, following both proper schedule and proper technique"*. Many of the existing adherence frameworks mainly focus on medication adherence and are centered around specific diseases and medication regimens [42].

These frameworks often draw upon health behavior models, including those exploring health information-seeking behavior [25, 31, 33], to understand and predict health-related behaviors. One such comprehensive model is the Information Motivation Behavioral Skills (IMB) model [18], which encompasses health-related information, motivation, and behavioral skills as key factors influencing behavior change (Fig. 1a). The IMB model has been successfully applied to various health-related behaviors, including medication adherence and breast self-examination behavior [18].

Another useful model is the generalized adherence loop model proposed by Klein et al. [28]. This model consists of three key nodes: Belief, Knowledge, and Action (Fig. 1b), and it aids in understanding the diverse factors influencing adherence behaviors. It serves as a framework for promoting adherence to a service or system and facilitating behavioral change. McLellan and Shelagh [37] emphasized the applicability of this model to preventive healthcare services, as it underscores the active involvement of individuals in managing and maintaining their own health.

However, despite the existence of these models, there remains a lack of comprehensive frameworks that address the multifaceted nature of adherence to health and fitness. This gap motivated our research to develop a new theoretical framework that extends beyond medication adherence and incorporates the broader context of health and fitness adherence among older adults.

Drawing upon the IMB and adherence loop models and integrating them with our study findings, we developed a theoretical framework as shown in Fig. 2. The framework consists of three levels. The nodes in the first level are taken from the IMB model [18] and the Adherence Loop model [28], as explained below. The second layer is derived from our study findings, highlighting the significance of information and the different types of knowledge. Finally, the third layer consists of the five

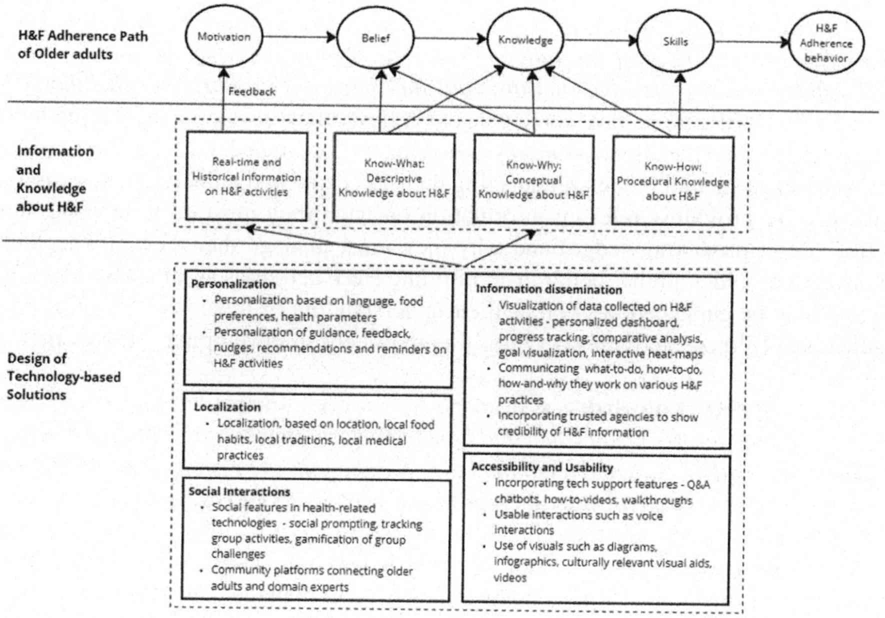

Fig. 2 Theoretical framework

key themes (personalization, localization, socailization, information dissemination, accessibility and usability) in designing technology-based solutions obtained from our study findings and subsequent design opportunities. The framework suggests that when designed effectively, technology-based solutions can facilitate access to information and knowledge about H&F for older adults, thereby promoting adherence to their H&F routines.

The first node, **Motivation**, is from the IMB model and is about *"drive or desire to engage in the behavior"*. According to the IMB model, motivation is crucial in influencing health behavior (adherence behavior in our study context). However, our study found that older adults often lacked adherence to a H&F routine despite showing initial motivation. This could be because of their limited knowledge regarding what needs to be done, leading to decreased motivation over time. Additionally, the IMB model proposes that motivation can influence behavior skills. The behavioral skills component of the IMB model focuses on an individual's objective and subjective abilities to perform a particular behavior. For example, suppose the behavior in question is exercising regularly. In that case, objective abilities include the knowledge of proper exercise techniques (skills), and subjective abilities include their perceived ability (self-efficacy) to exercise. While it requires appropriate knowledge to achieve objective abilities, motivation can influence an individual's sense of self-efficacy, which refers to their **Belief** (second node) in their ability to perform a particular behavior.

The second node, **Belief**, is from the Adherence Loop model, is about *"people's belief in the accuracy of the diagnosis, the appropriateness of the therapy, their self-efficacy to perform the therapy, and the therapy's validity and likelihood to succeed"*. Once people have some basis to believe, they develop a mental model of the condition and the effect of therapy on it. Our study revealed that many participants lacked a clear mental model of managing their H&F and had numerous concerns and queries. By providing relevant information on what they must do to manage their H&F (descriptive knowledge) and why they must manage their H&F (conceptual knowledge), older adults' belief in appropriate H&F activities as a means of healthy aging may be enhanced, as well as their understanding of the consequence of non-adherence to these activities. In turn, this may help in developing a robust mental model.

The third node, **knowledge**, is from the Adherence Loop model. The model states that *"a strong knowledge of when and how to perform the therapy will help a person develop the critical cues and reminders to stay adherent"*. Our study findings showed that older adults require different types of knowledge about H&F (descriptive, conceptual, and procedural). Providing relevant information on what older adults need to do to manage their H&F (descriptive knowledge), why they need to manage their H&F (conceptual knowledge), and how they need to manage their H&F (procedural knowledge) could enhance their knowledge and understanding of how to manage their H&F activities effectively.

The fourth node, **Skills**, is from the IMB model. As explained above, skills refer to an individual's objective abilities to perform a particular behavior, which can be achieved through appropriate knowledge and could promote **Adherence behavior**.

We propose a slight but significant modification to Klein et al.'s [28] model. They describe their model as a "loop", where individuals move through the three nodes of belief, knowledge, and actions to achieve adherence. They intend to create a loop of consistent interactions rather than a one-time linear engagement. While we agree that H&F activities are not a one-time engagement, we believe much will be accomplished in the initial phase of the H&F routine (unlike adherence to medication). Hence, we suggest an adherence "path" instead. Our proposed framework outlines a sequential journey that individuals undertake in their pursuit of adherence to H&F. The framework posits that individuals progress through these nodes in a linear fashion, with each step influencing the next. Although the framework follows a linear path, it incorporates essential feedback mechanisms. In the case of H&F routines, receiving feedback on one's actions (real-time and historical information on H&F) is more critical than the actions (or adherence behavior) themselves, as it reinforces motivation.

IMB model also says that well-informed and well-motivated individuals can effectively enact health behaviors. Real-time information on H&F informs individuals about their current H&F activities (*"Am I doing this exercise correctly?", "For how long have I been holding this pose?", or "Can I keep walking at this pace or do I need to slow down?"*), and historical information (*"Great, I have now been adherent to this diet for six months", or "Hmm, when I exercise, I sleep better"*) on H&F could motivate individuals to engage in these activities. All these factors can be readily supported by technologies available today, perhaps not by the designs in which these technologies are available.

Appropriately designed technology-based solutions provide relevant information and knowledge to older adults, fostering adherence to health and fitness regimens. For example, *group activity tracking* provides real-time and historical information on H&F activities of older adults in a group, serving as a motivating factor for adhering to a routine. Similarly, a *personalized Q&A chatbot*, tailored to an individual's health parameters, enriches their knowledge and skills, which in turn helps to adhere to a routine.

To conclude, our adherence framework proposes building technology-based solutions that can motivate older adults, enhance their belief, knowledge, and skills in H&F activities, and facilitate adherence behavior by providing relevant information and knowledge. By utilizing this framework, researchers can gain insights into the factors that positively influence adherence and utilize this knowledge to design interventions to improve health and fitness adherence among older adults.

6 Limitations and Conclusion

Although our study provides valuable insights, we recognize several limitations. First, our sample only includes older adults from educated, middle-class families living in urban or metropolitan areas of India. Therefore, the concerns and queries expressed by our participants may not represent other social groups, and our proposed solutions may not apply to those groups. Our findings are not intended to be generalized to the entire population of Indian older adults, but rather to offer an understanding of the experience of a specific subset of older adults who participated in our study. Second, as a qualitative study, our proposed solutions are provisional and subject to verification through future research. Finally, the proposed framework captures our current understanding of the context and identifies potential pathways to address the problem of non-adherence. While we acknowledge that there may be scenarios where the effectiveness of the framework is uncertain, pinpointing such instances without empirical study is challenging. The actual implementation, effectiveness, and efficacy of a specific design and the validation of the model are beyond the scope of this paper.

While technology may not be a panacea for all problems faced by older adults, it can assist them in adhering to H&F routines. Medical research and practice in India tend to focus on treating diseases and not much on aging or healthy living [17]. Therefore, there is a growing need for research in this area. By gaining a comprehensive understanding of the subject matter, our study provides valuable insights for designing technology-based solutions that promote adherence to H&F routines among Indian older adults.

Acknowledgements The authors are grateful to Dr Prabha Adhikari and Dr P. N. Manohar Herle for providing a comprehensive view of older adults' health and fitness scenario in India. The authors are also grateful to the older adults for participating in the interviews and to all the experts for their time answering our queries.

References

1. (n.d.) All about the HealthifyMe smart plan. Accessed 12 Sept 2022. https://www.healthifyme.com/blog/all-about-the-healthifyme-smart-plan/
2. (n.d.) Allopathic medicine. Accessed 15 June 2022. https://en.wikipedia.org/wiki/Allopathic_medicine
3. (n.d.) Ayurveda. Accessed 15 June 2022. https://en.wikipedia.org/wiki/Ayurveda
4. (n.d.) Classification of Indian cities. Accessed 12 Mar 2022. https://en.wikipedia.org/wiki/Classification_of_Indian_cities
5. (n.d.) Ghee. Accessed 15 June 2022. https://en.wikipedia.org/wiki/Ghee
6. (n.d.) Glycemic index. Accessed 20 Apr 2022. https://en.wikipedia.org/wiki/Glycemic_index
7. (n.d.) Liquorice. Accessed 15 June 2022. https://en.wikipedia.org/wiki/Liquorice
8. (n.d.) Pranayama. Accessed 15 June 2022. https://en.wikipedia.org/wiki/Pranayama
9. (n.d.) Tinospora cordifolia. Accessed 15 June 2022. https://en.wikipedia.org/wiki/Tinospora_cordifolia
10. (n.d.) Withania somnifera. Accessed 15 June 2022. https://en.wikipedia.org/wiki/Withania_somnifera
11. (2021) Ministry of Statistics and Programme Implementation. Accessed 25 Jan 2022. https://www.mospi.gov.in/documents
12. (2021) National Health Policy. Accessed 7 Feb 2022. https://nhm.gov.in/index1.php?lang=1&level=2&sublinkid=1046&lid=605
13. (2021) Stroke incidence and mortality: a report of the population based stroke registries, India. Accessed 21 Jan 2022. https://ncdirindia.org/All_Reports/pbsrbook/resources/PBSR_Report_2021.pdf
14. (2022) Smartphone users in India. Accessed 17 Apr 2023. https://www.start.io/audience/smartphone-users-in-india
15. Anand S (2022) India wearables market records double-digit growth in 1Q22, shipping 13.9 million units, says IDC India. Accessed 25 July 2022. https://www.idc.com/getdoc.jsp?containerId=prAP49123222
16. Borse S, Joshi M, Saggam A, Bhat V, Walia S, Marathe A, Sagar S, Chavan-Gautam P, Girme A, Hingorani L et al (2021) Ayurveda botanicals in COVID-19 management: an in silico multi-target approach. PLoS ONE 16(6):e0248479
17. Chawla G (2019) Healthy aging research in India. J Exp Res Hum Growth Aging 2
18. Fisher WA, Fisher JD, Harman J (2003) The information-motivation-behavioral skills model: a general social psychological approach to understanding and promoting health behavior. Soc Psychol Found Health Illness 22(4):82–106
19. Ghosh TS, Rampelli S, Jeffery IB, Santoro A, Neto M, Capri M, Giampieri E, Jennings A, Candela M, Turroni S et al (2020) Mediterranean diet intervention alters the gut microbiome in older people reducing frailty and improving health status: the NU-AGE 1-year dietary intervention across five European countries. Gut 69(7):1218–1228
20. Gibbs G (2008) Analyzing qualitative data. Sage
21. Harrington C, Martin-Hammond A, Bray KE (2022) Examining identity as a variable of health technology research for older adults: a systematic review. In: Proceedings of the 2022 CHI conference on human factors in computing systems (CHI'22), New Orleans, LA, USA. Association for Computing Machinery, New York, NY, USA, Article 265, 24 pp. https://doi.org/10.1145/3491102.3517621

22. Harrington CN, Wilcox L, Connelly K, Rogers W, Sanford J (2018) Designing health and fitness apps with older adults: examining the value of experience-based co-design. In: Proceedings of the 12th EAI international conference on pervasive computing technologies for healthcare, pp 15–24
23. Helbostad JL, Vereijken B, Becker C, Todd C, Taraldsen K, Pijnappels M, Aminian K, Mellone S (2017) Mobile health applications to promote active and healthy ageing. Sensors 17(3):622
24. Jennings A, Berendsen AM, De Groot LCPGM, Feskens EJM, Brzozowska A, Sicinska E, Pietruszka B, Meunier N, Caumon E, Malpuech-Brugère C et al (2019) Mediterranean-style diet improves systolic blood pressure and arterial stiffness in older adults: results of a 1-year European multi-center trial. Hypertension 73(3):578–586
25. Johnson JD, Meischke H (1993) A comprehensive model of cancer-related information seeking applied to magazines. Hum Commun Res 19(3):343–367
26. Kalimullah K, Sushmitha D (2017) Influence of design elements in mobile applications on user experience of elderly people. Procedia Comput Sci 113(2017):352–359
27. Kekade S, Hseieh MC-H, Islam M, Atique S, Khalfan AM, Li Y-C, Abdul SS (2018) The usefulness and actual use of wearable devices among the elderly population. Comput Methods Programs Biomed 153(2018):137–159
28. Klein DE, Wustrack G, Schwartz A (2006) Medication adherence: many conditions, a common problem. In: Proceedings of the human factors and ergonomics society annual meeting, vol 50. Sage, Los Angeles, CA, pp 1088–1092
29. Koh WK, Ng J, Tan O, Tay Z, Wong A, Helander MG (2009) Why taking medicine is a chore—an analysis of routine and contextual factors in the home. In: International conference on human centered design. Springer, pp 452–461
30. Kononova A, Li L, Kamp K, Bowen M, Rikard RV, Cotten S, Peng W et al (2019) The use of wearable activity trackers among older adults: focus group study of tracker perceptions, motivators, and barriers in the maintenance stage of behavior change. JMIR mHealth uHealth 7(4):e9832
31. Lambert SD, Loiselle CG (2007) Health information-seeking behavior. Qual Health Res 17(8):1006–1019
32. Laranjo L, Ding D, Heleno B, Kocaballi B, Quiroz JC, Tong HL, Chahwan B, Neves AL, Gabarron E, Dao KP et al (2021) Do smartphone applications and activity trackers increase physical activity in adults? Systematic review, meta-analysis and metaregression. Br J Sports Med 55(8):422–432
33. Lenz ER (1984) Information seeking: a component of client decisions and health behavior. Adv Nurs Sci 6(3):59–72
34. Li L, Peng W, Kononova A, Bowen M, Cotten SR (2020) Factors associated with older adults' long-term use of wearable activity trackers. Telemed e-Health 26(6):769–775
35. Li L, Peng W, Kononova A, Kamp K, Cotten S (2021) Rethinking wearable activity trackers as assistive technologies: a qualitative study on long-term use. In: Proceedings of the 54th Hawaii international conference on system sciences, p 3923
36. Maurya DK, Sharma D (2020) Evaluation of traditional ayurvedic preparation for prevention and management of the novel coronavirus (SARS-CoV-2) using molecular docking approach
37. McLellan S (2014) Precision medicine: the future of data-driven healthcare
38. Mendes R, Martins S, Fernandes L (2019) Adherence to medication, physical activity and diet in older adults with diabetes: its association with cognition, anxiety and depression. J Clin Med Res 11(8):583
39. Miura C, Chen S, Saiki S, Nakamura M, Yasuda K (2022) Assisting personalized healthcare of elderly people: developing a rule-based virtual caregiver system using mobile chatbot. Sensors 22(10):3829
40. Mohan V, Radhika G, Vijayalakshmi P, Sudha V (2010) Can the diabetes/cardiovascular disease epidemic in India be explained, at least in part, by excess refined grain (rice) intake? Indian J Med Res 131(3):369–373
41. Nayar S, Madhu SV (2020) Glycemic index of wheat and rice are similar when consumed as part of a North Indian mixed meal. Indian J Endocrinol Metab 24(3):251

42. Peh KQE, Kwan YH, Goh H, Ramchandani H, Phang JK, Lim ZY, Loh DHF, Østbye T, Blalock DV, Yoon S et al (2021) An adaptable framework for factors contributing to medication adherence: results from a systematic review of 102 conceptual frameworks. J Gen Intern Med 36(2021):2784–2795
43. Picorelli AMA, Pereira LSM, Pereira DS, Felício D, Sherrington C (2014) Adherence to exercise programs for older people is influenced by program characteristics and personal factors: a systematic review. J Physiother 60(3):151–156
44. Podder V, Nagarathna R, Anand A, Patil SS, Singh AK, Nagendra HR (2020) Physical activity patterns in India stratified by zones, age, region, BMI and implications for COVID-19: a nationwide study. Ann Neurosci 27(3–4):193
45. Rao P, Joshi A (2020) Design opportunities for supporting elderly in India in managing their health and fitness post-COVID-19. In: IndiaHCI'20: proceedings of the 11th Indian conference on human-computer interaction, pp 34–41
46. Rao Gadahad P, Joshi A (2022) Wearable activity trackers in managing routine health and fitness of Indian older adults: exploring barriers to usage. In: Nordic human-computer interaction conference (NordiCHI'22), Aarhus, Denmark. Association for Computing Machinery, New York, NY, USA, Article 7, 11 pp. https://doi.org/10.1145/3546155.3546645
47. Rastogi A (2018) Physical activity. Accessed 21 Jan 2022. https://www.nhp.gov.in/healthlyliving/physical-activity
48. Rhodes RE, Martin AD, Taunton JE, Rhodes EC, Donnelly M, Elliot J (1999) Factors associated with exercise adherence among older adults. Sports Med 28(6):397–411
49. Rivera-Torres S, Fahey TD, Rivera MA (2019) Adherence to exercise programs in older adults: informative report. Gerontol Geriatr Med 5(2019):2333721418823604
50. Ryu H, Kim S, Kim D, Han S, Lee K, Kang Y (2020) Simple and steady interactions win the healthy mentality: designing a chatbot service for the elderly. In: Proceedings of the ACM on human-computer interaction, CSCW2, vol 4, pp 1–25
51. Saggam A, Limgaokar K, Borse S, Chavan-Gautam P, Dixit S, Tillu G, Patwardhan B (2021) Withania somnifera (L.) Dunal: opportunity for clinical repurposing in COVID-19 management. Front Pharmacol 835
52. Silva PA, Holden K, Aska N (2014) Smartphones, smart seniors, but not-so-smart apps: a heuristic evaluation of fitness apps. In: Foundations of augmented cognition. Advancing human performance and decision-making through adaptive systems. Lecture notes in computer science, vol 8534. Springer, Cham
53. Sivaraju S, Moneer A, Gangadharan KR, Syamala T, Supriya V, Nidhi G (2017) Caring for our elders: early responses India ageing report—2017. Accessed 21 Feb 2022. https://india.unfpa.org/sites/default/files/pub-pdf/India%20Ageing%20Report%20-%202017%20%28Final%20Version%29.pdf
54. Sullivan AN, Lachman ME (2017) Behavior change with fitness technology in sedentary adults: a review of the evidence for increasing physical activity. Front Public Health 4(2017):289
55. Sun S (2022) Number of smartphone users in India in 2010 to 2020, with estimates until 2040. Accessed 11 July 2022. https://www.statista.com/statistics/467163/forecast-of-smartphone-users-in-india/
56. Talukder MS, Sorwar G, Bao Y, Ahmed JU, Palash MAS (2020) Predicting antecedents of wearable healthcare technology acceptance by elderly: a combined SEM-neural network approach. Technol Forecast Soc Change 150(2020):119793
57. Teixeira E, Fonseca H, Diniz-Sousa F, Veras L, Boppre G, Oliveira J, Pinto D, Alves AJ, Barbosa A, Mendes R et al (2021) Wearable devices for physical activity and healthcare monitoring in elderly people: a critical review. Geriatrics 6(2):38
58. Vargemidis D, Gerling K, Spiel K, Abeele VV, Geurts L (2020) Wearable physical activity tracking systems for older adults—a systematic review. ACM Trans Comput Healthcare 1(4):37 pp, Article 25. https://doi.org/10.1145/3402523
59. Victor CR, Rogers A, Woodcock A, Beighton C, Cook DG, Kerry SM, Iliffe S, Whincup P, Ussher M, Harris TJ (2016) What factors support older people to increase their physical activity levels? An exploratory analysis of the experiences of PACE-lift trial participants. Arch Gerontol Geriatr 67 (2016)

60. Wang S, Bolling K, Mao W, Reichstadt J, Jeste D, Kim H-C, Nebeker C (2019) Technology to support aging in place: older adults' perspectives. In: Healthcare, vol 7. Multidisciplinary Digital Publishing Institute, p 60
61. Waterworth J, Waterworth E, Álvarez P, Gutiérrez J, Carús JL, Garcia S (2016) What do elderly users want and need from fitness technologies?: findings from the ELF@ home project. In: Integrating technology in positive psychology practice. IGI Global, pp 104–126
62. Watson S (2022) Millet for diabetes: how it affects blood sugar. Accessed 10 July 2022. https://www.webmd.com/diabetes/millet-diabetes-blood-sugar
63. WHO (2010) Global recommendations on physical activity for health. World Health Organization, Geneva, p 60

Design and Evaluation of a Password Diversifier Tool

Rekha Pathak, Anjali Aggarwal, R. Aiswarya, Jayasree Raveendran, Manish Shukla, Vijayanand Banahatti, and Sachin Lodha

Abstract Passwords can be well regarded as an integral aspect of our lives. However, reuse of passwords is prominent, making various accounts vulnerable to attacks. Within reuse, password structural reuse is a highly prevalent issue. Password structure is an ordered sequence that captures password composition using four character classes: Lowercase (L), Uppercase (U), Number (N), and Special Symbols (S). In case of targeted password guessing attack on a user, the risk of password disclosure increases manifold when the user reuses the password structure. To discourage reuse and enhance awareness of risk due to similarity in password structures, we test whether feedback to users on structural reuse risk would result in creation of diversified passwords. We develop an initial version of a tool named Password Diversifier

Conference acronym 'XX, Jan XX-XX, 20XX, Place, City

ACM ISBN 978-1-4503-XXXX-X/18/06…$15.00
https://doi.org/XXXXXXX.XXXXXXX.

R. Pathak · A. Aggarwal · R. Aiswarya (✉) · J. Raveendran · M. Shukla · V. Banahatti · S. Lodha
TCS Research, Pune, India
e-mail: aiswarya.r1@tcs.com

R. Pathak
e-mail: rekha.pathak@tcs.com

J. Raveendran
e-mail: jayasree.raveendran@tcs.com

M. Shukla
e-mail: mani.shukla@tcs.com

V. Banahatti
e-mail: vijayanand.banahatti@tcs.com

S. Lodha
e-mail: sachin.lodha@tcs.com

A. Joshi and G. R. Sim (eds.), *Proceedings of the 14th Indian Conference on Human-Computer Interaction*, Studies in Computational Intelligence 1162,
https://doi.org/10.1007/978-981-97-4335-3_3

and analyze password behavior in an experimental setup where users could enter various passwords or their structures and make changes to these based on recommendations. The tool analyses password diversity through similarity metrics, by comparing user's password structures against each other as well as against some predefined heuristics. The tool also nudges the user to improve their password structures through in-built recommendations towards safer structures. We find from an experimental study with 53 users that users are able to create diversified passwords based on password risk related information given to them.

Keywords Password structure · Password reuse · Password diversifier · Nudges

CCS Concepts: • **Security and privacy → Human and societal aspects of security and privacy**

1 Introduction

Despite advancements in the realm of user security such as biometric authentication and face-recognition systems, text-based passwords continue to be the most common mode of authentication. Lack of awareness and preference for convenience drive users to adopt password behaviors which can prove to be highly risky [31]. An example is the creation of text-based passwords using common words and placements of uppercase, lowercase, special symbols and digits in [27, 42].

One major undesirable security behavior that most users are unaware of is the reuse of passwords or the use of slightly modified passwords for multiple sites. For instance, Stobert and Biddle [35] found that 96% of interviewees in their study reused passwords. The main reason for users to adopt this behavior is a preference for convenience, as reuse would not require them to remember several passwords. For example, users do not prefer securing their social media and shopping accounts as much as they prefer securing their bank accounts. The trade-off between convenience and security must be given attention to [10].

However, some children are excluded from playing most games, not because of lack of access to resources, but because of a physical or a mental disability, which renders their access impossible. Since a lot of games rely heavily on visual cues, they are inaccessible to the visually disabled.

Password diversity refers to having different passwords for different accounts. In the past, researchers have looked at password diversity at organization level [33] and developed a feedback system that helps a user in selecting a password with a permitted character class structure (pattern of special symbols, digits, uppercase and lowercase letters). We think the following three concerns may arise with this approach:

(1) Reduced password space—NIST special publication on digital identity guidelines [12] mandates eight characters as the minimum length requirement for

password when all four character classes (special symbols, digits, uppercase and lowercase letters) are used. However, the guideline recommends fewer character-classes for passwords with large length (usually 12 or more). Based on this strategy, an eighth length password with all four character classes mandated will yield $4^8 = 65{,}536$ potential structures [19, 21]. For a two-character class passphrase with a length of eighteen characters, the total number of potential password structures will be $2^{18} = 262{,}144$. This is a small space for large organizations where the employee count is more than 500,000 people. Hence, this approach will reduce potential password creation space considering people will use the minimum length recommended by password policy.

(2) Usability concern—If password diversity is implemented at organization level [33], it will create issues for users. The organization disables a particular password structure after it has been seen to be used beyond a threshold value. That is, once a set of users use a particular password structure, it will make that particular structure unusable for other users. For new users using the same structure will mark that structure as "already used" and will not allow password creation attempts using that structure. This reduction in password creation space will force users to pick another structure leading to usability concern.

(3) Bad password practice—Past research has shown that the user tends to adopt bad password practices, often referred to as coping strategies, when they are faced with usability issues [13]. These coping strategies range from use of personal information, well-known patterns and slight modification to the existing password patterns.

In this work we present a user-specific password diversifier, which is reasonable due to two primary reasons. First, it helps in making a targeted password guessing attack on a user more difficult as our proposed solution detects and responds to reuse of structures. Second, an attacker requires only a single credential to break into multiple accounts if the user reuses passwords. Therefore, it is imperative to deter the attack at user level, which is possible when the password structures are diversified.

Password structure is an ordered sequence that captures password composition using four character classes: Lowercase (L), Uppercase (U), Number (N), and Special Symbols (S). Each password structure has multiple password creation possibilities. Consider a 4-character-class set (L = Lowercase letters a–z, U = Uppercase letters A–Z, D = Decimal numbers 0–9, S = Special symbols), which is used to create a password of length "n". All these password structures are called password bins. Every bin represents a class of passwords, e.g., bin L8 represents 8 length passwords composed entirely of lowercase letters while the bin U1L7 represents 8 length passwords that begin with an uppercase letter followed by 7 lowercase letters. Thus, the password space is divided into a total of c^n bins, where c is the number of character classes available for password creation. Illustration of a bin space for password length of 2 with 4 alphabet set usage is shown here in Fig. 1.

Some users use password managers to store passwords thus reducing cognitive load [2]. While password managers are available, their use is fairly limited [3]. The enterprise users have a bigger problem given the need to remember passwords of their

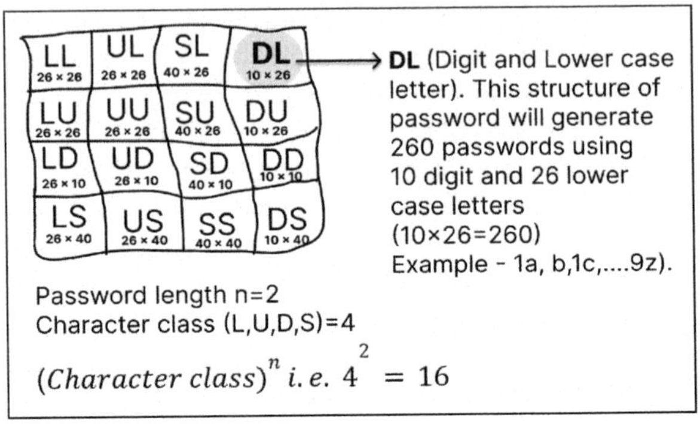

Fig. 1 Password space for the given password structure

work accounts in addition to personal accounts. The tendency to reuse passwords is further increased as users have to create many passwords over time due to expiration policies associated with these multiple accounts. While servers store one-way hash of passwords which will be used to check similarity, even a slight change in a password character like using ‘1’ instead of ‘3’ will create a different hash making it appear dissimilar but without addressing issues in actual diversity. Hence, there is a need to compare passwords using structures. In a study on adaptive password-composition policies, the author observed significant increase in password strength with little to no decrease in usability if structure-based password policy is used [33]. This idea motivated us to use password structures in our analysis to assist users to analyze their own passwords set, which is more in the user’s control.

In the current work, we analyze password structure-based reuse risk through a password diversifier tool in an experimental setting. The tool we have developed help users to diversify the structures of their passwords using similarity metrics for checking the password structures against each other as well as against some predefined blacklisted heuristics [39]. Users could either directly enter their passwords or the password structures. Apart from analyzing all the password structures for similarity and common blacklisted heuristics, the tool also recommends new distinct structures to diversify the existing ones. This paper discusses the design of the tool and results of an experimental study done to evaluate the tool.

2 Related Work

Text-based passwords are still the most common and primary authentication method for securing our accounts. Due to lack of awareness and memorability issues, most users tend to reuse their passwords or use a partially modified version of existing passwords for most of their online services [9, 15, 30, 40]. Even after a service gets breached, many users tend not to change their reused passwords for a long time [40] or change to a similar or weak password [7]. Studies also show that frequently entered passwords tend to be more reused across most of the websites [41]. Wash et al. [41] found that apart from more frequently used passwords, more complex passwords also tend to be reused. The most prominent reason cited for reusing passwords is the difficulty of remembering many passwords [9, 28].

Other reasons cited include having too many accounts, accounts not being important enough to warrant a new password, the ease of using same password for the same category of accounts and so on [28, 41]. Hanamsagar et al. [14] find security fatigue, that is, the point when maintenance of online security becomes too exhausting or burdensome to the user as a major reason for password reuse. Misconceptions about security also prompt password reuse [14]. One study [1] looks at behavioral determinants of password reuse in work and personal spaces found that people who reuse passwords scored significantly high on attributes such as neutralization (tendency to justify undesirable behavior), impulsivity (lack of self-control) and neuroticism (lack of emotional stability). Apart from reuse, users tend to use common structures of passwords which can be easily guessed and attacked. Password reuse has been mostly addressed through mitigation strategies such as notifications to reset password [11], mnemonic techniques, graphical passwords, token-based or smart card authentication, or the use of biometrics and multiple step authentication [16].

Password managers have evolved significantly providing better support to users in terms of password management and security by providing features like password generator, auto-fill and evaluating password health in addition to secure storage of passwords, yet little research has been done in motivating users to adopt such applications [25]. Users tend not to use password managers due to lack of awareness or uncertainty of the security of such applications [31]. In some cases, inability to evaluate one's own vulnerability discourages individuals from adopting them [4]. Users who use browser-based or third-party password managers tend not to use additional security features like password generators due to memorability and cross-device usage issues [29]. Many of the popular password managers are coming up with innovative solutions to reduce password breaches by introducing a security interface. Recently, LastPass introduced a security dashboard which measures password security and reuse. It also includes dark web monitoring to notify the user when a breach of their information or password has occurred [5, 23]. 1Password also has an insights dashboard which features breach checks and password health metrics for the user to evaluate their security behavior [26]. Despite such efforts users are still reusing their passwords across websites due to lack of awareness about the presence of such functionality or effective usage of them [24, 31].

Much work has been done to encourage users to create strong passwords in the form of text feedback and blacklists [13, 17, 33, 36, 38, 39]. Using blacklists has proven effective in creating strong passwords only if configured carefully. The wordlist and matching algorithm can have a significant effect on improving strength and usability of passwords [36]. A structure-based blacklist has been found to have a more pronounced effect on increasing the password strength than a string-based blacklist [33]. It has also been found that giving detailed feedback can lead users to create more secure, and no less memorable passwords than using only a bar to show the strength [13, 39]. To calculate the strength of the password, many researchers have proposed using common password heuristics [17, 38, 39]. When passwords are created in the presence of such heuristics, they tend to be more secure and diverse and create awareness among users about good password creation practices [17]. While these requirements affect the strength of passwords created, it is necessary to investigate the way these requirements are communicated to the users. These findings will also help the third-party password managers in incorporating these requirements in their tools to motivate users in creating strong and diverse passwords [34]. Prior research in evaluating password strength and diversity and analyzing user behavior in password management shows that users store their passwords as readable text which exposes them to high risk of breaches and often discourage users in entering their real passwords [6]. This opens up the need for a more secure way of analyzing passwords which users can trust and can give a clear picture of their password management behavior.

Based on Thaler and Sunstein's idea of nudging [37], several previous studies in the cybersecurity domain have found subtle persuasive techniques to be effective in enhancing security behavior. For instance, through an intra-subject experiment where different nudges were administered to see different levels of comfort with retaining auto-generated passwords, the authors found that salience nudge was most effective in reducing this comfort level [18]. Another study focusing on user choice between open and secure networks used color gradient and order of the networks to make most secure ones appear on the top of the list in more prominent colors [8]. Applying this concept, the tool nudges the user by highlighting risk using a color code (red) and highlighting 'Change Password' button in the tool. This is further facilitated by giving recommendations for better password structure. The recommendations are introduced as a choice, where users can use the recommended structure to create their own new structures when changing the password. While information is presented in a salient way, there is also a choice architecture created, where the user could choose to use a system recommended password structure or use one's own structure. Thus, the user is also endowed with agency to take informed decisions for creating diversified passwords.

Given this evidence of rampant reuse of passwords and limited use of password managers, addressing the problem at a structural level is recommended. Reuse of passwords tend to follow the particular structure (for instance, uppercase first and special symbol at the end) since users usually only slightly modify their passwords. Thus, addressing password reuse at a structure level will lead to creation of more unique passwords. Unlike at the organization level, when we urge users to diversify passwords at user level, it could enhance use of unique passwords through diversified structures. We developed the Password Diversifier, which uses structure-based checks on password structures to give users information about the similarity with their other password structures and common heuristic (blacklist) structures. Apart from analyses of password structure, the tool also nudges the user to change a password when structure similarity is found to be high.

3 Password Diversifier: Concept and Features

Users generally deal with a number of password-protected work and personal accounts. In the light of the challenges discussed above, there is a need for a tool that can effectively inform users about risks due to structural similarities in their passwords across all accounts and nudge them to change undesirable structures. Users may not realize or understand the intensity of risk unless it is presented to them in a clearly comprehensible format. Moreover, even if a user realizes the presence of risk in passwords, it would be very difficult for them to manually create a password structure which passes checks against bad heuristics (blacklist) and does not belong to leaked password structure database in addition to being dissimilar to the structures of their other existing passwords. So, the tool, which we call 'Password Diversifier' has been developed with features such as diversity analysis, presentation of information on structure-based risk, nudges for the user to change password structures when high risk is identified and a choice architecture design to create structurally dissimilar passwords to mitigate risk. The objective of the tool is to urge users to change structurally similar passwords that are reused across accounts.

Password Diversifier uses password structures and not the actual passwords for analysis. The tool does not store actual passwords of users. The reason which guided this decision was the need to avoid capturing password text which may lead to potential leakage scenario, while providing the scope for diversification. Action logs of users capture insights into user's usage of the tool. The Password Diversifier is a standalone tool in its existing initial version, which does not need internet and it saves only structures of password for analysis which can be verified by users in log files.

3.1 Features of the Tool

Password Diversifier was designed to help users to measure similarity of password structures entered and verify if a password structure fails heuristic checks. The following sections describe the computations involved.

3.1.1 Similarity Ratio Computation for Password Structures and Heuristics Structures

The first function of the tool is to compute similarity ratio for password structures. Similarity ratio is based on Levenshtein distance [22] which is a string metric for measuring the difference between two sequences. It is like any other statistical metric which is used commonly in computer science field. Informally, the Levenshtein distance between two words is the minimum number of single-character edits (insertions, deletions, or substitutions) required to change one word into the other. It is

Table 1 Example of password structure similarity ratio (S-ratio) and heuristic ratios (H-ratio)

Login and password structure	Other login sites	Password structure	S-ratio	Overall S-ratio	Failed heuristic checks for the given login (out of 17)	H-ratio
Gmail (ULLLLLL-LLSNNNN)	OffVPN	ULLLLLLLSNNN	0.92	0.78	ULLLLLLLSNNN (H1), ULLLLLLLSNNN (H6),	0.29
	Webmail	ULLLLLLLSNNN	0.92		ULLLLLLLSNNN (H8),	
	Bank1	ULLLULLN	0.63		ULLLLLLLLSNNNN (H14),	
	Bank2	ULLLULLN	0.63		ULLLLLLLSNNN (H16)	
OffVPN (ULLLLL-LLSNNN)	Gmail	ULLLLLLLLSNNNN	0.92	0.83	ULLLLLLLSNNN (H1),	0.24
	Webmail	ULLLLLLLSNNN	1.00		ULLLLLLLSNNN (H6),	
	Bank1	ULLLULLN	0.70		ULLLLLLLSNNN (H8),	
	Bank2	ULLLULLN	0.70		ULLLLLLLSNNN (H16)	

named after the Soviet mathematician Vladimir Levenshtein, who considered this distance in 1965. Levenshtein distance ratio ranges between 0 and 1. 0 means that two structures are not at all similar and 1 means they are perfectly similar.

The overall similarity ratio (S-ratio) of any password structure is the mean of all its similarity ratios with other individual password structures as shown in Table 1. Computation of similarity ratio is a two-step process: (a) For all password structures entered in the tool, each password's individual similarity ratio with every other password structure is computed and (b) overall similarity ratio which is the mean of all its similarity ratios with other individual password structures is arrived at. This ratio helps to understand how similar a particular password structure is to other password structures. More diverse password structures will have the ratio values closer to 0.

The system also assesses every password structure for blacklist heuristic checks [17, 38, 39]. This includes checks for common blacklisted patterns like number at the end or upper case in the beginning. A previous study by [39] found heuristics to be effective in scoring password strength and generating text-based feedback for users. Ur [38] identifies 21 heuristics from which we took 17 which were applicable for a structure-based password. The 17 blacklist heuristic checks chosen are stated in Table 2. These password heuristics are used to compute Heuristics ratio (failed heuristics count/total heuristic count) as shown in Table 2.

3.1.2 Visual Representations of Analysis Results and Risks

When there is a perfect match between password structures or when a structure is failing on some blacklist related heuristic checks, then the cell in the table containing a corresponding report link gets highlighted in yellow. Clicking on any such alerts

Table 2 Blacklist heuristic checks

H1 = Password should not contain 2 or more consecutive numbers	H10 = Password does not contain lower case letter though it is allowed
H2 = Password should not contain 2 or more consecutive upper-case letters	H11 = Password does not contain upper case letter though it is allowed
H3 = Password should not contain 2 or more consecutive special symbols	H12 = Password does not contain number though it is allowed
H4 = Password should not have special symbol just at the end	H13 = Password does not contain special symbols though it is allowed
H5 = Password should not contain number at the beginning	H14 = Password may contain date pattern MMMM + YYYY
H6 = Password should not contain number at the end	H15 = Password may contain date pattern MMMM + YY
H7 = Password should not contain all numbers	H16 = Password with letter-special symbol-number pattern is not recommended
H8 = Password should not contain uppercase character at the beginning	H17 = Password must not be the part of chosen common blacklisted structure list
H9 = Password should not contain all uppercase characters	

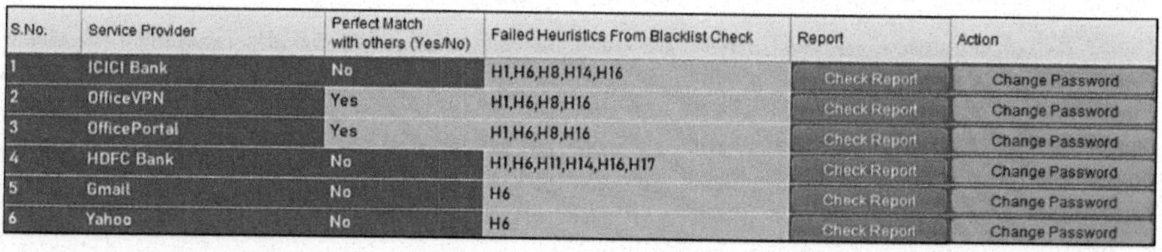

S.No.	Service Provider	Perfect Match with others (Yes/No)	Failed Heuristics From Blacklist Check	Report	Action
1	ICICI Bank	No	H1,H6,H8,H14,H16	Check Report	Change Password
2	OfficeVPN	Yes	H1,H6,H8,H16	Check Report	Change Password
3	OfficePortal	Yes	H1,H6,H8,H16	Check Report	Change Password
4	HDFC Bank	No	H1,H6,H11,H14,H16,H17	Check Report	Change Password
5	Gmail	No	H6	Check Report	Change Password
6	Yahoo	No	H6	Check Report	Change Password

Fig. 2 Depiction of similarity of password structures and failed heuristics

will display a detailed report of the password structure on the source of risk and suggests the user to change password which appears as a button highlighted in red color (Fig. 2).

The Password Diversifier displays the results of analyses and generates alerts in different visual and color-coded formats that are easily understandable and appealing to the user. The tool provides two views. In matrix view, the similarity ratios of all password structures are presented in a matrix format for easy and complete comparison (Fig. 3).

In graphical view, each password protected account is plotted in a graph setting using the similarity and blacklist ratios calculated from their corresponding scores (Fig. 4). In this view, the user can see the diversity status of its password structure portfolio. Points are plotted for each password account considering its similarity ratio and blacklist ratio. The background of graphic view is colored with green-amber-red depiction enabling the user to view the password with reference to its risk status (green indicating least risk and red indicating highest risk). All well-diversified structures will be in the green zone.

Personal
Work

Gmail	Gmail					
OfficeVPN	61.538464	OfficeVPN				
OfficePo...	61.538464	100.0	OfficePo...			
ICICI Bank	77.41936	82.75862	82.75862	ICICI Bank		
HDFC Bank	66.666664	88.0	88.0	86.666664	HDFC Bank	
Yahoo	86.666664	64.28571	64.28571	72.72727	62.068966	Yahoo

Fig. 3 Matrix view of similarity ratios

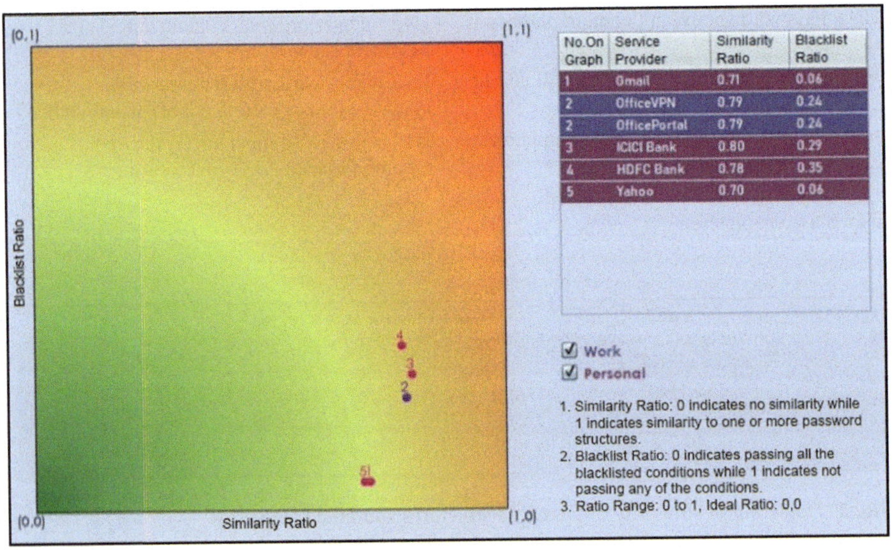

Fig. 4 Graphical view of similarity ratios and blacklist ratios

When users find a password structure to be risky, they can click the ‘Change Password’ button which would be highlighted in red. This built-in nudge alerts the user and suggests changing the password with a given set of recommended structures (Fig. 5).

The user can go back and forth in the tool to make changes to the password multiple times to reduce the risk score. Once user sets the new password, reanalysis is done (Fig. 6). If the user follows any recommendation suggested by the system and sets the new password accordingly, then that service passes all the diversity and blacklist check criteria.

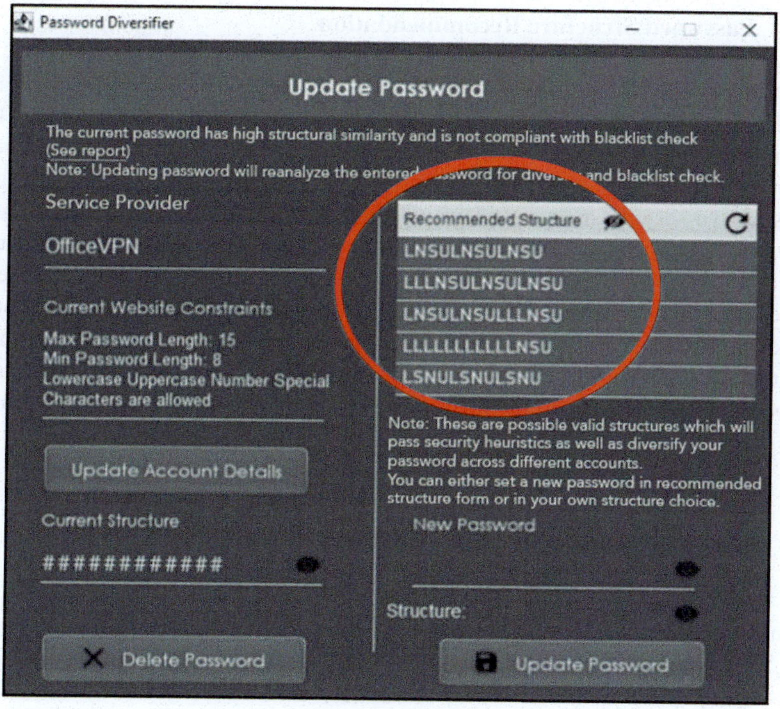

Fig. 5 Pop-up after user clicks on 'Change Password'

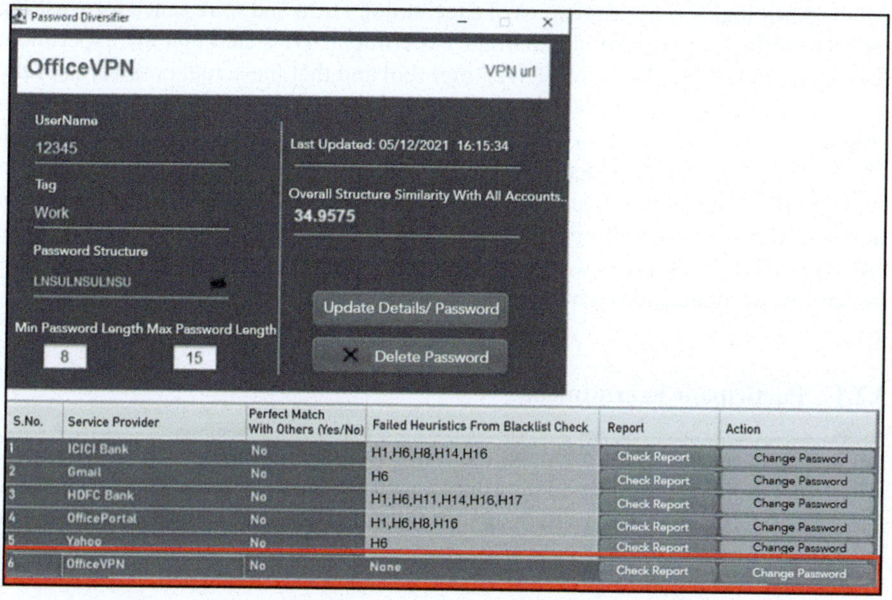

Fig. 6 Depiction of results of reanalysis

3.1.3 Password Structure Recommendation

When there is 100% similarity ratio for any password structure entered, or if any structure is failing on heuristics check, then that structure is labelled as 'High-Risk'. All high-risk passwords are provided with recommendations for changing the corresponding password structures. In such a recommended structure, the constraints of respective service sites are taken into consideration. For e.g., if number is not allowed in passwords for a particular service site, then the recommended structure choices for that service will not contain numbers. From the available character categories, the tool constructs structures that do not match other password structures entered in the tool. Thus, overall similarity is reduced, and it is ensured that the recommendation does not contain any of the 17 blacklisted heuristics. The tool suggests a maximum of 5 such recommended structures at a time, which can be refreshed for more recommendations till the user finds a structure of their choice.

3.2 Experimental Setup

To analyze the effectiveness of Password Diversifier, we conducted an experiment with 53 volunteers. We aimed at a sample size of greater than 30 participants, as per the normally accepted criterion for conducting parametric tests in statistics [20]. The experiment was preceded and succeeded by a survey. The pre-experiment survey collected data on the number of passwords handled by users in work and personal spaces. All the participants were employees of a technology firm and were sent a form asking for their willingness to participate in the experiment. We stated that the experiment consisted of using the Password Diversifier tool and that the actual passwords of the users would not be saved anywhere. The compensation for participation based on the existing reward system for such activities in the organization was communicated. The participants were provided compensation equivalent of 5 dollars after successful completion of the experiment. Thus, the experiment had voluntary participation of subjects who were clearly given a briefing of the objective of the experiment and tasks expected. Participants were encouraged to enter the actual passwords they used for various accounts although it was not a mandatory activity.

3.2.1 Participant Recruitment

A mail was sent out to all the people in the firm with all the details of the experiment, including the process and the compensation provided. The mail also enclosed a form which the interested participants could fill, giving their consent to participate in the experiment. All the details were communicated clearly to all the participants before commencing the experiment. The experiment was conducted over a duration of 45 days with a batch of around 10 participants in one session. Every participant was allowed to use the tool for three days and we completed the experiment with 53 participants over 5 sessions. Of the 53 participants who completed the experiment, 52 consented to giving their demographic information. Of these, 28.8% of participants identified as female and the rest as male. 71.1% of participants were between 18 and 34 years of age and the rest were between 45 and 54 years. 48% of the participants had a tenure with the organization between 0 and 5 years and the rest had a tenure above 5 years. In the three-days duration, participants can enter their passwords from various websites, check the analysis results, and change their passwords according to the recommendations provided. All participants in the experiment received a nominal compensation for their time and effort, as per organizational guidelines on compensating efforts for taking part in such research activities with necessary approvals.

3.2.2 Participant Instructions and Pre-experiment Survey

We shared step-by-step instructions on how to use the tool with our participants in both written and video format. Participants were first directed to take the pre-experiment survey. Using a 5-point Likert scale, responses were obtained on how the passwords were stored and their risk perception related to passwords. Self-reported measures were obtained for the number of password-protected accounts users have in work and personal categories, the likelihood of these passwords being vulnerable to a security threat, the scope for guessing other passwords of the user if one of the passwords were guessed right and so on (see Appendix). Once the survey was completed, we shared the written document of the instructions along with the executable file of the tool. This was followed by an online meeting with the participants where we played a demonstration video of downloading and using the tool. The purpose for an online meeting with the participants was to clarify any concerns of the participants if they had any. Also, the participants could clear their doubts about the steps for using the tool and how the tool was created. Participants were then asked to sign up on the tool and start using it. It was made clear to participants that changing the password would be activated only within the tool meaning that the tool does not actually connect with any external service account site for enabling change of passwords. This was to ensure that users felt safe using the tool. The tool was entrusted with them for a maximum period of three days, and they were asked to share action log files once they were done using the tool. Logs were captured during the window of the experiment. Logs contained information details of password structures entered, various ratios required for data analysis and user actions along with timestamp.

3.2.3 Using the Tool

Once participants signed up on the tool, they were asked to add as many accounts credentials as they maintain. The various fields to be filled included Name of the account or website (Example, LinkedIn), Website Constraints involved with the particular website, Website Category (Work or Personal) and Password Structure. Users can opt either to enter the actual password or convert the password into its structure form and enter it in the password structure field. If the user opts to enter a password, then the tool converts the password directly to its structure form and saves that structure for future reference and analysis purposes. The actual passwords are never stored in the database.

Initial passwords structures entered by users can be analyzed by them using the 'Analyze' button in the tool. On completion of this action, they are presented with the buttons for two reports—Similarity check and Heuristic check. In the similarity check report, they can view the similarity scores of a particular password structure with other password structures entered by them. In the Heuristic check report, they can view the heuristics which have failed for that password structure. Once participants find that the password structure is risky, the button 'Change Password' is highlighted in red color, alerting the user of the risk and the need to change the password. On the click of 'Change Password' button, five different recommendations for new password structure options would be shown, from which they could select any one to make their existing password diverse. The user can also choose to create totally new structures of their own wish.

The entire process of use of tool and data capture is depicted in Fig. 7. At the start, the user has to sign-up in the system by giving his/her email id and setting some password for the system. There is a help document which the user can access at any time. The user needs to provide details about various services getting used day by day where password is required. Those services can be related to work, like domain access password, company internal portal password etc. or it can be of personal use like email services, shopping services, banking services etc. The user needs to enter details of these services like service provider name (Gmail/ICICI/Amazon etc.), URL, username on those sites, password constraints which that service site has (minimum/maximum length, special symbols allowed etc.) and the actual password. The system does not store this actual password, but it converts it to structure format and then saves structure in the system. User can add/update/delete such services details. On every add/update/delete, the system analyzes all entered password structures at backend. Python module is getting called to compute Levenshtein distance ratio of each structure with other structures. This analysis is done at various levels: considering all structures, considering only work-related structures, considering all personal category structures, and finally considering each individual subcategory in personal passwords (email, banking, shopping, news, and entertainment). Blacklist heuristic check module is getting called to validate added/modified structure against 17 blacklisted structure criteria.

The user can access this analysis result in various ways. Main analysis result screen displays a table (Fig. 2) with one row for each service password structure. It shows if that structure matches perfectly with any other structure (yes/no). If yes, that cell is highlighted in yellow color. In another column, it displays all failed blacklist heuristic check numbers, if any (also in yellow color). There is a button to check a detailed report about failed blacklist heuristics and similarity ratio with each other individual structure. If any of these two columns in main analysis result table has warning in yellow colors, then last column has a button asking to change password and it is highlighted in pink color. When user clicks on that "Change Password" button, the

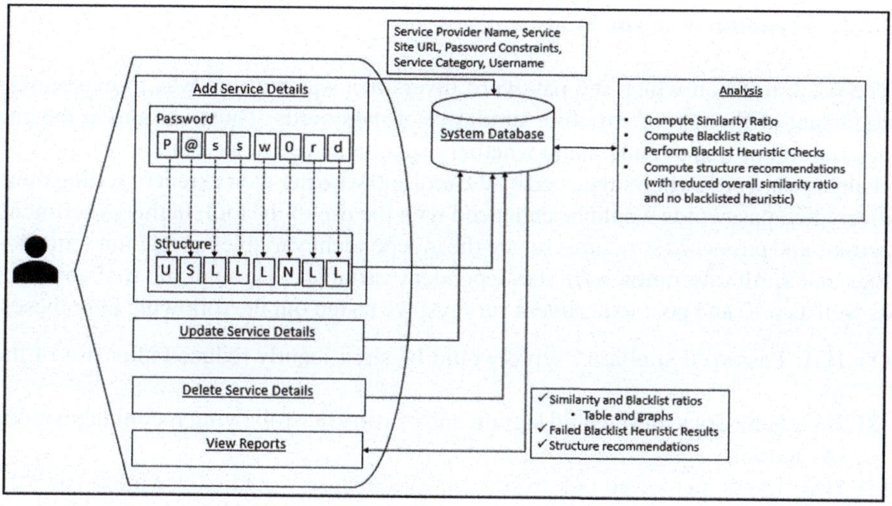

Fig. 7 Process of the password diversifier tool

system provides 5 new structures as a recommendation for new password (Fig. 5). These structures are already checked for diversification with existing other structures and for blacklisted criteria. Also, these structures adhere to service site constraints for password. If user sets new password in one of these recommended forms, then it automatically clears all the analysis tests, and no warning comes for that structure. User can click on refresh button to get more recommendations, if not satisfied with provided set of 5. User can see similarity ratio of each password structure with other password structures in matrix form also (Fig. 3). User can view overall status of all password structures in a graph view also where x axis shows similarity ratio and y axis shows blacklist ratio (Fig. 4). This graphical view can quickly tell user whether current password structures are well diversified and blacklist-proof. Both the ratios 0 is ideal situation and both the ratios 1 is worst situation. For all user actions, logs are generated in JSON format. After updating the password structure, analysis of results gets automatically repopulated for all the accounts in the tool.

3.2.4 Post-experiment Survey

After the experiment, another survey was administered to collect users' perceptions of the tool's utility aspects. These included ratings on the perceived importance of the tool, impact on awareness of password diversification, ease of use, confidence to manage password security and perceived usefulness of the recommendations and analytics.

3.2.5 Hypotheses of the Study

The basic need on which the password diversifier was built on was to make users cognizant of the risk of structure similarity of passwords. Thus, our aim at the end of experiment was to understand whether users make use of the features of the tool, change passwords when suggested by the tool and whether their extent of setting more diversified passwords would be enhanced with the use of the tool. In this experiment, actions and password structures set are the independent variables and action sequence logs and similarity ratios were the dependent variables. Through the tool analytics as well as pre- and post-experiment surveys, we tested out the following hypotheses:

(1) Ha1: Password similarity score would be significantly reduced after use of the tool.
(2) Ha2: Heuristics ratio would significantly drop after following recommendations of the tool.
(3) Ha3: Users' perceived risk from password misuse would reduce after using the tool.
(4) Ha4: Different age groups will have different perceived importance of the tool after using it.
(5) Ha5: Different gender groups will have different perceived importance of the tool after using it.
(6) Ha6: Different tenure groups will have different perceived importance of the tool after using it.
(7) Ha7: There will be a difference in perceived risk from potential threats to passwords between different age groups.
(8) Ha8: There will be difference in perceived risk from potential threats to passwords between different gender groups.
(9) Ha9: There will be difference in perceived risk from potential threats to passwords between different tenure groups.

Gender and tenure were investigated, since these demographic characteristics have shown differences in password behavior in previous studies. For instance, Petrie and Merdenyan [32] find that men fared worse in password recall than women. With regard to tenure, Aiswarya et al. [1] find that senior employees have higher risk perception of password related threats and that for work passwords, women tend to reuse lesser compared to men.

3.2.6 Addressing Privacy Concerns

Password Diversifier is designed and developed as a standalone tool with features described in Sect. 3.1. It can be installed on windows desktop and does not need internet. Since it is a standalone tool, the password structures would be saved locally. As stated earlier, it saves only the structures of password and not the actual password. The users also had access to the log files and complete transparency relating to the use of logs for analysis was maintained. They could access the log files if they wished to.

4 Analysis and Insights

4.1 Activities with Passwords in the Tool

The tool provided the choice to either enter a password or the corresponding structure. It was observed that users entered passwords as against structures 69.02% of times. Users entered passwords related to both work and personal accounts. 43.66% of the passwords were work-related and 56.34% were passwords in the personal space. The largest number of total passwords entered by users is 29 in this study. The maximum number of work-related passwords entered was 10 and the corresponding number for personal accounts was 23. On average, users were found to have entered a greater number of personal accounts than work related accounts.

4.2 Password Risk Analysis

Initially when password structures were entered by participants, an average of 40.85% of passwords had high risk because of the presence of a 100% similarity of structures when compared with other password structures. The top three most used password structures across users were ULLLSNNN, ULLLLSNNNN, ULLLLLSNNNN. Similarly, out of all the password structures initially, an average of 95.13% passwords had high risk because of failure with reference to heuristic checks. Heuristic checks which failed for the maximum number of times were H1 ("Password should not contain 2 or more consecutive numbers"), H8 ("Password should not contain uppercase character at the beginning"), H6 ("Password should not contain number at the end") and one which failed the least number of times is H17 ("Password must not be the part of chosen common blacklisted structure list").

After the experiment had ended and the user had changed their passwords based on the suggestions of the tool, the similarity scores of password structures were again analyzed. We find that risk scores of password structures significantly reduced for both work and personal accounts, thus rendering evidence that users engaged in diversifying their passwords in both work and personal accounts. For work accounts, t-test analysis of similarity scores pre-test (M = 0.80, SD = 0.02) and post-test (M = 0.64, SD = 0.01) indicate a significant difference, $t(50) = 8.07, p = 0.001$. Similarly, we notice a statistically significant reduction in similarity scores of password structures for the personal accounts as well (Pre-test (M = 0.73, SD = 0.02) and post-test (M = 0.63, SD = 0.01); $t(50) = 4.47, p = 0.001$). Thus, our hypothesis that similarity scores of password structures significantly reduces after use of tool (Ha1) is confirmed.

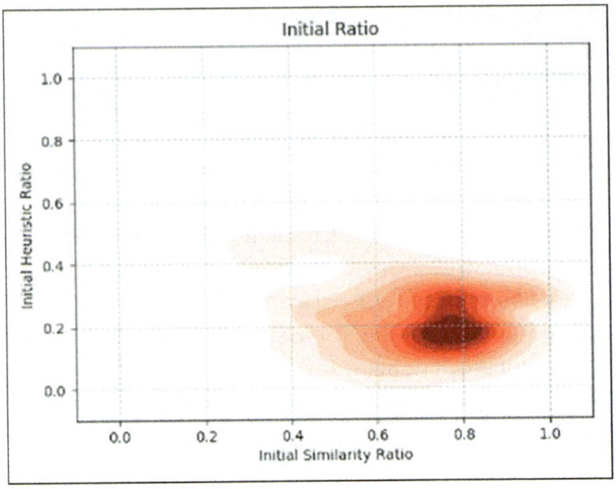

Fig. 8 Graphical representation of initial similarity and heuristics ratio

4.3 User Adoption of Recommendations from the Tool

It is found that 23.50% users followed the recommendation of the tool (of which 42.73% pertained to work category and 57.27% to personal accounts). The rest did not follow the recommendation and set the password with structures of their choice. The most important outcome is that in all the instances of password change, the structural similarity was reduced to zero when recommendation was followed. When the tool recommended password structure was not followed, reduction of similarity risk is by 93.71%. Thus, even when the tool recommendation is not adopted by the user, acting on the 'Change password' suggestion has also reduced the similarity risk score. We used the Kernel Density Estimation for data smoothing based on a finite data sample [38]. The scatter plots (Figs. 8 and 9) created using this statistical approach illustrate the reduction in ratios which is depicted by shift towards bottom part of the graph. This suggests that there is an improvement towards desirable password structures at the end of experiment, with reduced scores on heuristic ratios and similarity ratios. Initially the heuristic ratios for passwords entered were higher but after changing passwords and following recommendations, the number of failed heuristics reduced thus reducing the ratio. We find a significant reduction in the heuristics ratio before ($M = 0.21$, $SD = 0.008$) and after the experiment ($M = 0.13$, $SD = 0.016$), as confirmed by a t-test ($t(552) = 12.79$, $p < 0.01$) Ha2 is thus supported, meaning that heuristics ratio significantly drops after use of the tool. However, reduction in similarity ratios were not as good as reduction in heuristic ratios, owing to users' structure reuse habits persisting.

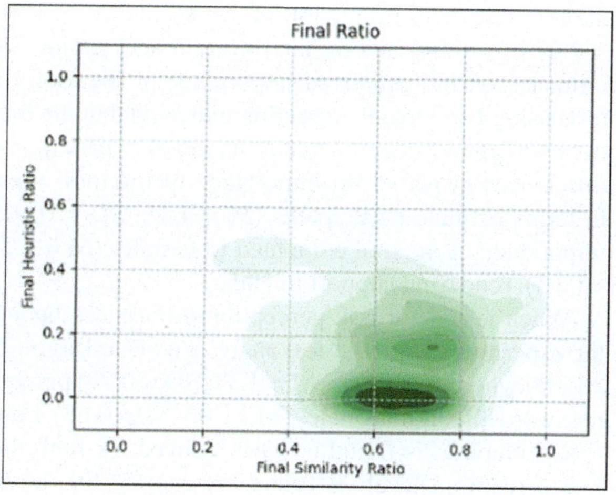

Fig. 9 Graphical representation of final similarity and heuristics ratio

4.4 *Perceived Risk of Password Misuse*

Perceived risk of password misuse was measured in the pre-experiment survey using a 5-point Likert scale on two dimensions: susceptibility of their existing passwords to a threat and the likelihood that other passwords of the user could be guessed if one of their passwords were guessed correctly. Participants in general had a neutral score (M = 3.05, SD = 0.73) indicating a lack of surety about their susceptibility to threat and guessability of their passwords. However, post the experiment, their risk perception of password misuse significantly reduced after usage of the tool (M = 1.49, SD = 0.66) as confirmed by a t-test ($t(50) = 10.65$; $p < 0.01$). Thus, users were more confident about their new passwords being free from threat. Even if one of their passwords gets compromised in a security breach, remaining passwords will remain safe due to diversity. Thus, perceived risk from passwords is seen to significantly decline after use of the tool, rendering support to Ha3.

4.5 *Insights from Demographic Analyses*

We analyzed the responses of participants on tool effectiveness on parameters viz. importance of the tool, awareness of password security, ease of use, increased confidence in creating secure passwords, usefulness of analytics, usefulness of recommendations and ability to implement of the tool recommendations. Demographic variables considered in the study were age group, gender and experience at work. Analysis of variance was conducted for various sub-groups based on age classes (18–24; 25–34; 35–44 and 45–54) and experience classes (Less than 1 year; 1–5 years, 5.1–10 years, 10.1–20 years and more than 20 years). The results of ANOVA indicate a statistically significant difference emerging for the perceived importance of the tool for better protection of passwords in work and personal accounts ($F(47,3) = 2.92$, $p = 0.04$) among the different age groups. We find that age group had an influence on the perceived importance of the tool, thus confirming Ha4. Post-hoc tests using Bonferroni correction indicated that the importance of the tool was found to be the greatest for the age group 35–44 years ($p < 0.03$). When tested for gender differences in perceived importance of the tool, t-test further revealed differences in terms of male participants (M = 4.45; SD = 0.60) scoring higher on perceived importance of the tool compared to females (M = 4.2; SD = 0.56, $t(50) = -1.42$, $p < 0.05$), rendering support to Ha5.

When tested for risk perception differences between males and females before the experiment started, t-test analysis revealed significant difference ($t(50) = 2.36$; $p < 0.01$) wherein females (M = 3.33; SD = 0.79) perceived greater risk from potential password threats than males (M = 2.83; SD = 0.63). Post the tool use in the experiment, risk perception is found to have reduced for both the genders, though the drop in risk perception levels is found to be more for the male cohort (M = 1.43; SD = 0.60) compared to females (M = 1.73; SD = 0.70) ($t(50) = 1.55$, $p < 0.05$). Thus, as hypothesized (Ha8), we find significant differences in risk perception between genders. We do not find evidence confirming Ha6, Ha7 and Ha9.

5 Limitations and Future Work

In this experiment we use an initial version of a standalone tool where the user has to manually change the updated passwords on the website, and we currently have no means to check whether such change has been made in the real passwords that users have. The standalone nature of the tool prevented the passwords being linked to the corresponding website and the option of autofill being made available to the users. The next version of the tool could make this feature available, while ensuring that safeguards to prevent privacy and data breach concerns are intact.

Though the effectiveness of the recommended structure is proven, more refinements may be needed to enhance the usability features such as enabling users to remember the changed structures better. The recommendations given by the tool could be improved to make minimal changes to an existing password while satisfying the criteria for a structurally diversified password. This feature is currently not incorporated due to the inherent challenges in creating such recommendations that satisfy all the 17 heuristic criteria, for which a common algorithm needs to be conceived. In the current version of the tool, we have not tested for the strength and memorability of the passwords with diversified structures.

It should also be noted that the password changes carried out by users took place in an experimental setting within a fixed time duration, which was about three days per participant. Studies involving longer durations of time can enable a more nuanced understanding of user behavior in terms of implementing the recommendations of the tool and ensuring that password structure risks are mitigated.

Most of the users participating in the experiment came up with their own structures to reduce the similarity and heuristic ratio. This could be because the recommendations were randomly chosen which passed all the heuristic and similarity checks. We are trying to formulate better recommendations which are more suited to each user as a part of our future work.

6 Discussion and Conclusion

In this paper, we discuss the development of a Password Diversifier tool to help users diversify their passwords. Three key features of the tool are: firstly, the tool compares password structures (with other password structures entered) and computes a similarity score. Secondly, heuristic assessment for every password structure is also conducted. Thirdly, users are alerted of risks and nudged to change passwords, facilitated by recommendation of password structures We conducted an experiment with 53 users, who were given a choice to enter either their passwords or password structures belonging to work and personal accounts. Based on activities in the tool and their analyses, we find that the tool is well acknowledged by participants, and it is proven to be effective in reducing similarity ratio and improving the structural diversity of password structures. By the end of the experiment, the similarity and heuristic scores have reduced with a more significant reduction in heuristic score. This resulted largely when the users acted upon the Change Password nudge given by the tool. We observe that there is a 100% reduction of risk score when users have followed the recommendations of the tool for creating structurally dissimilar passwords. The perceived risk of password misuse has also come down significantly as indicated by post-experiment survey responses of the participants.

Appendix

Pre-experiment Survey Questions

(1) How do you usually store your passwords? (Open ended question)
(2) How many password-protected work accounts do you have? If you are unable to recall, an approximate number would be great! (Open ended question)
(3) How many personal email and social media accounts do you have? (Open ended question)
(4) How many financial accounts do you have? (Open ended question)
(5) How many media, entertainment, ticket-booking and shopping accounts do you have? (Open ended question)
(6) How likely do you think your password-protected accounts are susceptible to a security threat? (Not at all likely/Very unlikely/Neutral/Very likely/Extremely likely)

(7) If one of your passwords is guessed right, how likely is it that other passwords can be guessed? (Not at all likely/Very unlikely/Neutral/Very likely/Extremely likely)
(8) Would you be able to create a password that is difficult to hack if you had guidance on how to create one? (Most definitely yes/There's a good chance I might/Maybe/I don't think it would make a difference to me)
(9) What are some barriers that you face in dealing with passwords in general? (Open ended question)

Post-experiment Survey Questions

(1) I feel this tool is important for better security of password protected work and personal accounts (Strongly Agree/Agree/Neutral/Disagree/Strongly Disagree)
(2) I am now more aware of the risk of password similarity after using this tool (Strongly Agree/Agree/Neutral/Disagree/Strongly Disagree)
(3) The tool has been easy to use (Strongly Agree/Agree/Neutral/Disagree/Strongly Disagree)
(4) I feel more confident about the security of my accounts after making changes based on risk highlighted by the tool (Strongly Agree/Agree/Neutral/Disagree/Strongly Disagree)
(5) I think that analytics on password diversification is needed for me (Strongly Agree/Agree/Neutral/Disagree/Strongly Disagree)
(6) I find the recommendation of password structures by the tool to be useful (Strongly Agree/Agree/Neutral/Disagree/Strongly Disagree)
(7) I was able to modify my password structures based on recommendations by the tool (Strongly Agree/Agree/Neutral/Disagree/Strongly Disagree)
(8) Would you prefer this tool to store passwords (not simply password structures) and serve as a password manager as well? (Yes/No/Maybe)
(9) Would you prefer a mobile version of this tool? (Yes/No/Maybe and accepted open-ended feedback)
(10) Which age group do you belong to? (18–24/25–34/35–44/45–54/55–64/65 and above/Do not wish to disclose)

References

1. Aiswarya JR, Banahatti V (2022) Behavioral attributes in password reuse: analysis of password practices in work and personal spaces. In: Proceedings of the 13th Indian conference on human-computer interaction
2. Alodhyani F, Theodorakopoulos G, Reinecke P (2020) Password managers—it's all about trust and transparency. Future Internet 12(11):189
3. Aurigemma S, Mattson T, Leonard L (2017) So much promise, so little use: what is stopping home end-users from using password manager applications?

4. Ayyagari R, Lim J, Hoxha O (2019) Why do not we use password managers? A study on the intention to use password managers. Contemp Manag Res 15(4):227–245
5. Bachmann L (2020) What are dark web scans? Accessed 2 Mar 2023. https://blog.lastpass.com/2020/08/what-are-dark-web-scans/
6. Berardi D, Callegati F, Melis A, Prandini M (2020) Password similarity using probabilistic data structures. J Cybersecur Priv 1(1):78–92
7. Bhagavatula S, Bauer L, Kapadia A (2020) (How) Do people change their passwords after a breach? arXiv preprint arXiv:2010.09853
8. Briggs DJLCP, van Moorsel A (2014) Nudging whom how: it proficiency, impulse control and secure behaviour. Networks 49(2014):18
9. Erickson J (2011) Online authentication: password reuse and security questions. In: The 11th Winona computer science undergraduate research symposium, p 10
10. Fagan M, Khan MMH (2016) Why do they do what they do?: a study of what motivates users to (not) follow computer security advice. In: Twelfth symposium on usable privacy and security (SOUPS 2016), pp 59–75
11. Golla M, Wei M, Hainline J, Filipe L, Dürmuth M, Redmiles E, Ur B (2018) "What was that site doing with my Facebook password?" Designing password-reuse notifications. In: Proceedings of the 2018 ACM SIGSAC conference on computer and communications security, pp 1549–1566
12. Grassi P, Newton E, Perlner R, Regenscheid A, Burr W, Richer J, Lefkovitz N, Danker J, Choong Y-Y, Greene K, Theofanos M (2017) Digital identity guidelines: authentication and lifecycle management. https://doi.org/10.6028/NIST.SP.800-63b
13. Habib H, Colnago J, Melicher W, Ur B, Segreti S, Bauer L, Christin N, Cranor L (2017) Password creation in the presence of blacklists. Proc USEC 2017:50
14. Hanamsagar A, Woo S, Kanich C, Mirkovic J (2016) How users choose and reuse passwords. Information Sciences Institute
15. Hanamsagar A, Woo SS, Kanich C, Mirkovic J (2018) Leveraging semantic transformation to investigate password habits and their causes. In: Proceedings of the 2018 CHI conference on human factors in computing systems, pp 1–12
16. Hoonakker P, Bornoe N, Carayon P (2009) Password authentication from a human factors perspective: results of a survey among end-users. In: Proceedings of the human factors and ergonomics society annual meeting, vol 53. Sage, Los Angeles, CA, pp 459–463
17. Jayakrishnan GC, Sirigireddy GR, Vaddepalli S, Banahatti V, Lodha SP, Pandit SS (2020) Passworld: a serious game to promote password awareness and diversity in an enterprise. In: Sixteenth symposium on usable privacy and security (SOUPS 2020), pp 1–18
18. Kankane S, DiRusso C, Buckley C (2018) Can we nudge users toward better password management? An initial study. In: Extended abstracts of the 2018 CHI conference on human factors in computing systems, pp 1–6
19. KoreLogic (2015) PAM module and library for auditing/enforcing password topology histogram wear-leveling. Accessed 2 Mar 2023. https://github.com/KoreLogicSecurity/libpathwell
20. Kwak SG, Kim JH (2017) Central limit theorem: the cornerstone of modern statistics. Korean J Anesthesiol 70(2):144–156
21. Leininger H (2015) LibPathWell 0.6.1. Accessed 2 Mar 2023. https://blog.korelogic.com/blog/2015/07/31/libpathwell-0_6_1
22. Levenstein VI (1966) Binary codes with correction of loss, insertion and substitution of symbols. Sov Phys Dokl 10(1966):707–710
23. Lu K (2020) LastPass's new security dashboard includes dark web monitoring. Accessed 2 Mar 2023. https://identityreview.com/lastpass-new-security-dashboard-includes-dark-web-monitoring/
24. Lyastani SG, Schilling M, Fahl S, Backes M, Bugiel S (2018) Better managed than memorized? Studying the impact of managers on password strength and reuse. In: 27th USENIX security symposium (USENIX security 18), pp 203–220

25. Maclean R, Ophoff J (2018) Determining key factors that lead to the adoption of password managers. In: 2018 international conference on intelligent and innovative computing applications (ICONIC). IEEE, pp 1–7
26. Merkulov P (2022) Announcing insights from 1Password. Accessed 2 Mar 2023. https://blog.1password.com/announcing-insights/
27. Morris R, Thompson K (1979) Password security: a case history. Commun ACM 22(11):594–597
28. Notoatmodjo G, Thomborson CD (2009) Passwords and perceptions. In: AISC, vol 9. Citeseer, pp 71–78
29. Oesch T (2021) An analysis of modern password manager security and usage on desktop and mobile devices
30. Pearman S, Thomas J, Naeini PE, Habib H, Bauer L, Christin N, Cranor LF, Egelman S, Forget A (2017) Let's go in for a closer look: observing passwords in their natural habitat. In: Proceedings of the 2017 ACM SIGSAC conference on computer and communications security, pp 295–310
31. Pearman S, Zhang SA, Bauer L, Christin N, Cranor LF (2019) Why people (don't) use password managers effectively. In: Fifteenth symposium on usable privacy and security (SOUPS 2019), pp 319–338
32. Petrie H, Merdenyan B (2016) Cultural and gender differences in password behaviors: evidence from China, Turkey and the UK. In: Proceedings of the 9th Nordic conference on human-computer interaction, pp 1–10
33. Segreti SM, Melicher W, Komanduri S, Melicher D, Shay R, Ur B, Bauer L, Christin N, Cranor LF, Mazurek ML (2017) Diversify to survive: making passwords stronger with adaptive policies. In: Thirteenth symposium on usable privacy and security (SOUPS 2017), pp 1–12
34. Shay R, Bauer L, Christin N, Cranor LF, Forget A, Komanduri S, Mazurek ML, Melicher W, Segreti SM, Ur B (2015) A spoonful of sugar? The impact of guidance and feedback on password-creation behavior. In: Proceedings of the 33rd annual ACM conference on human factors in computing systems, pp 2903–2912
35. Stobert E, Biddle R (2014) The password life cycle: user behaviour in managing passwords. In: 10th symposium on usable privacy and security (SOUPS 2014), pp 243–255
36. Tan J, Bauer L, Christin N, Cranor LF (2020) Practical recommendations for stronger, more usable passwords combining minimum-strength, minimum-length, and blocklist requirements. In: Proceedings of the 2020 ACM SIGSAC conference on computer and communications security, pp 1407–1426
37. Thaler RH, Sunstein CR (2009) Nudge: improving decisions about health, wealth, and happiness. Penguin
38. Ur B (2018) SIGCHI outstanding dissertation award–supporting password decisions with data. In: Extended abstracts of the 2018 CHI conference on human factors in computing systems, pp 1–3
39. Ur B, Alfieri F, Aung M, Bauer L, Christin N, Colnago J, Cranor LF, Dixon H, Naeini PE, Habib H et al (2017) Design and evaluation of a data-driven password meter. In: Proceedings of the 2017 chi conference on human factors in computing systems, pp 3775–3786
40. Wang C, Jan STK, Hu H, Bossart D, Wang G (2018) The next domino to fall: empirical analysis of user passwords across online services. In: Proceedings of the eighth ACM conference on data and application security and privacy, pp 196–203
41. Wash R, Rader E, Berman R, Wellmer Z (2016) Understanding password choices: how frequently entered passwords are re-used across websites. In: Twelfth symposium on usable privacy and security (SOUPS 2016), pp 175–188
42. Weir M, Aggarwal S, De Medeiros B, Glodek B (2009) Password cracking using probabilistic context-free grammars. In: 2009 30th IEEE symposium on security and privacy. IEEE, pp 391–405

Graphical Passwords for Emergent Users: A Four-Day Recall Comparative Study on PIN, Passfaces and Celebrities

Asif P Zuha, Anirudha Joshi, Sanika Deshpande, Anniket Mohalik, Rohit Naik, and Vijayanand Banahatti

Abstract Numerical passwords (or PINs) are a commonly used "what you know" password mechanism. However, numbers could be difficult to remember, and particularly for emergent users. Graphical passwords, and especially Passfaces (passwords based on human faces) could be an alternative, as we humans remember and recognise other human faces easily. Further, we posit that if these human faces happen to be of well-known personalities (such as celebrities), these should be even easier to remember. In this paper, we present a within-subjects, counterbalanced, longitudinal study with 66 emergent users that systematically compares the performance of PINs with Passfaces of two types, one composed of unknown people (Passfaces for short) and Passfaces based on celebrities (Celebrities for short). In our study, Passfaces and Celebrities had a larger password space by design (and hence, arguably, were more secure) than PINs. Users were asked to memorise each password, and recall the same after a distraction task and after a period of 4 days. The analysis of the results showed that the 4-day memorability of PINs (90%) and Celebreties (93%) were both significantly better than Passfaces (77%). While people took significantly more time to set

A. P. Zuha (✉) · A. Joshi · S. Deshpande · A. Mohalik
Indian Institute of Technology, Bombay, Mumbai, Maharashtra, India 400076
e-mail: zuhaasif.work@gmail.com

A. Joshi
e-mail: anirudhajoshi@iitb.ac.in

S. Deshpande
e-mail: sanikad96@gmail.com

R. Naik
Symbiosis International University, Pune , Pune, Maharashtra, India

V. Banahatti
TCS Research, Pune , Pune, Maharashtra, India

A. Joshi and G. R. Sim (eds.), *Proceedings of the 14th Indian Conference on Human-Computer Interaction*, Studies in Computational Intelligence 1162,
https://doi.org/10.1007/978-981-97-4335-3_4

up a Celebrities password (180 s) compared to PIN (101 s) and Passfaces (138 s), they took significantly more time to recall a Passfaces password (87 s) compared to PIN (42 s) and Celebrities (45 s). We present an analysis of the security and memorability of the faces used in Passfaces and Celebrities. We discuss the implications of our results on the design of password mechanisms for emergent users.

Keywords Emergent users · Graphical passwords · Privacy and security

CCS Concepts:• **Human-centered computing** → **Usability testing**; • **Security and privacy**→ *Domain-specific security and privacy architectures*; **Usability in security and privacy**; • **Social and professional topics** → **Seniors**

1 Introduction

As technology products such as smart phones and the internet permeate the society more widely, they are used more widely and by a diverse group of users, and for diverse tasks. Researchers have referred to several user groups from developing countries as "emergent" users. Such user groups may be less educated, economically disadvantaged, older, geographically dispersed, or have a culturally heterogeneous background (Fig. 1) [7].

Knowledge-based authentication (also known as "what you know" password) is the most popular method for a user to identify themselves [9]. This is commonly text-based. Numerical passwords, or personal identification numbers (PINs) are typically composed of a particular number of digits and do not entail any letters or symbols. However, PINs tend to have a limited security space and are not considered to be suitable for all applications (e.g. internet banking). The "stronger" what-you-know passwords, are alphanumerical, and may contain a combination of (Latin) text, symbols and numbers. However, alphanumeric passwords have a major disadvantage. Because human memory is limited, most users tend to choose simple or short passwords that are easy to remember [15]. This could potentially lead to security issues.

Fig. 1 Numerical password, Passfaces with unknown faces and Passfaces with celebrity faces

It has been argued that emergent users find it difficult to use text-input mechanisms [7]. Creating and recalling a robust alphanumeric password can be an arduous task even for individuals who are well-educated and familiar with technology. It is a much more pronounced challenge for emergent users who may have low levels of literacy and (in non-Latin-script countries) are usually less familiar with reading or typing in the Latin script. Unfortunately, the current what-you-know password systems do not account for the unique challenges faced by this user group. Nonetheless, ensuring privacy and security remains a critical concern.

What-you-know passwords could also be graphical. These are a type of knowledge-based authentication that attempt to leverage human memory for visual information, which arguably is less dependent on language or the ability of the user to read or write. The popular pattern lock [1] in current Android phones is the most common example of graphical passwords, though (like the PINs) these too tend to have a limited security space.

Though it is less common, passwords could also use photographs in different ways. One such example is Passfaces proposed by Brostoff and Sasse, which is an interaction technique wherein the user recognise a series of human faces from a grid to authenticate themselves [3]. A study compared Passfaces to traditional passwords in a field trial involving 34 participants over three months [3]. Passfaces had fewer login errors than passwords, even when the intervals between logins were long. If the number of possible images is large enough, the password space of a graphical password technique could be comparable with, or even surpass alphanumeric passwords. Therefore, the use of graphical passwords might offer better resistance to different security attacks [10]. (We provide a more detailed review of graphical passwords below.)

In this paper, we are motivated to compare the performance of emergent users with PINs and Passfaces. Further, we hypothesise that using faces of known people in the Passfaces interface might be more beneficial for memorability. To evaluate this hypothesis, we compare two versions of Passfaces – one with unknown faces (called Passfaces for short) and one with well-known faces (called Celebrities for short). Thus, through a within-subjects counterbalanced study with emergent users (N = 66) using three interaction techniques (PIN, Passfaces and Celebrities) for setting a 6-digit password, we make the following contributions:

- We propose a lab-based method to rigorously compare interaction techniques for setting passwords in a relatively short period of time.
- We report successes of first-time-usability, 4-day memorability, and a combined effect on the success of first-time-usability and 4-day memorability of the three techniques. Through our studies, we demonstrate that it is possible to design password mechanisms that can be used by emergent users and recalled after 4 days with above 80% success rates.
- We analyse the popularity of the celebrities in our chosen set, and report on the relationship between popularity and memorability while used for Passfaces.
- We present an analysis of the number of attempts and the percentage of success and failures at each attempt.

2 Background

While alphanumeric passwords are currently the gold standard, the security provided by alphanumeric passwords has been an area of concern. The types of passwords that are most easily cracked are the account holder's name, words from the dictionary, numbers related to personal information, permuted words, and a combination of all of these [2]. Alphanumeric passwords can be written down and stored as well. A hacker may also attempt to logically associate the date of birth and other numbers related to personal information as their first guess.

In addition to logical guessing, there are other types of attacks that can compromise password security, such as shoulder surfing. This type of attack involves a malicious user skillfully observing another user as they enter their password [16]. PIN based passwords are common in mobile device authentication due to ease of use with respect to space constraint, but it has issues with shoulder surfing and smudge attacks. To mitigate this risk, the layout of the password entry interface can be randomized, but participants of a study conducted by Kirkwood et al. [12] were quite averse to using a randomized keyboard, and most will not do so, unless forced by the system [13] conducted a comprehensive analysis of the security of smartphone unlock PINs. The study was based on the analysis of over 80,000 smartphone PINs collected from participants through a user study. The study concludes with recommendations for improving PIN security, including the use of longer and more complex PINs. The study analyzed the distribution and strength of PINs, the impact of different factors such as demographics and device type on PIN selection, and the effectiveness of different PIN guessing attacks. The results of the study show that a significant number of users select weak and predictable PINs and that factors such as age, gender, and device type have a significant impact on PIN selection. The study also found that PIN guessing attacks can be highly effective, with over 40 percent success rate in some cases. The study concludes with recommendations for improving PIN security, including the use of longer and more complex PINs, and the implementation of mechanisms to prevent brute-force attacks (e.g. limiting the number of retries). In our study, a password length of 6 instead of 4 was considered in order to increase the complexity as mentioned in the above paper.

2.1 Graphical Passwords

A systematic literature review [16] identifies four types of graphical passwords – recognition-based, recall-based, cued recall-based and hybrid. Recognition-based methods, such as Cognometrics/Search Metrics requires users to select specific images during password creation and later identify them from a set of other images during authentication. Passfaces are an example of recognition-based graphical passwords [16]. Recall-based methods involve users drawing or reproducing a pre-defined drawing on a grid to authenticate. The pattern lock in current Android phones is an example of this method. Cued recall-based methods, also known as Locimetric Schemes, prompt users to select password click points in a specific order within an image, for example: select a certain section of an image in a particular order. While conceptually similar to recall-based passwords, these mechanisms provide additional (often visual) cues to aid recall. Graphical Password Authentication Using Cued Click Points: A Usability Analysis by Biddle, Chiasson, and Van Oorschot [4], found CCP to be more memorable and efficient, offering an intuitive way for users to authenticate themselves. The usability analysis reveals several advantages of CCP compared to traditional alphanumeric passwords. The graphical nature of CCP allows users to visually identify click points on a selected image, making it easier to remember and reproduce their password. This advantage contributes to improved user experience and potentially reduces the risk of forgotten passwords or the need to write them down. In another paper "PassPoints" by Wiedenbeck et al. [19], demonstrates that PassPoints offers enhanced usability and security. Users were able to successfully authenticate using PassPoints over a password-based system. The system's usability benefits contribute to improved user satisfaction and potentially reduce the risk of password-related issues, such as forgotten or easily guessed passwords. This indicates the usefulness of a graphical password system.

Lastly, hybrid graphical passwords combine multiple graphical password schemes to leverage their respective strengths, creating a more robust authentication method. CD-GPS is a good example of a hybrid graphical password which is a click-draw-based scheme proposed by Meng [14]. It is a combination of cued-recall and recall methods that utilises elements of the PassPoint [19], CCP [4], Story [5], and DAS [11] schemes.

In this paper, we explore Passfaces proposed by Brostoff and Sasse [3], which is a recognition-based password mechanism. Instead of requiring users to reproduce complex drawings or sequential clicks, recognition-based graphical passwords leverage the user's visual recognition skills, arguably making them more intuitive and accessible for individuals who may have limited experience with technology or literacy challenges. In Passfaces, users recognise pre-selected faces from a grid to authenticate themselves, and multiple authentication rounds are used. Passfaces prompts users to select a series of faces from a grid of pictures. As mentioned previously, a study [3] compared Passfaces to traditional passwords in a field trial involving 34 student participants over three months. Passfaces had fewer login errors than passwords, even when the intervals between logins were long.

De Angeli et al. investigated the usability and security of pictorial passwords compared to PINs in their paper "Usability and User Authentication: Pictorial passwords vs. PIN" [6]. They found that humans struggle to remember meaningless strings, and after a week's interval, pictures were less error-prone than numbers without compromising the speed of the transaction. Participants reacted positively to the Visual Identification Protocol (VIP) concept, perceiving it as more secure and easier to remember than PINs. However, the study also revealed that VIP's usability can be easily disrupted by inappropriate solutions, and the results need to take into account the novelty factors of VIP. The paper compares picture passwords such as flowers, food, etc. against pins. VIP was found to be easier to remember and preferred by users, but its usability can be easily disrupted by an inappropriate solution.

Thorpe and Van Oorschot [17] investigate user choice in graphical passwords in their paper "An Empirical Investigation of User Choice in Graphical Passwords". The study highlights the advantages of user-centered design in graphical password systems. By analyzing data from a large-scale online survey, the authors identify the factors influencing users' choices of graphical passwords. The findings reveal that users prefer graphical passwords with certain characteristics, such as familiarity, simplicity, and ease of memorization. Understanding these user preferences allows for the design of graphical password systems that align with users' mental models and increase their overall satisfaction and engagement with the authentication process.

In the paper "The Emperor's New Security Mechanism: A Graphical Password Generator" by Dunphy [8], the advantages of a graphical password generator based on user-selected images are discussed. The proposed approach offers usability and security benefits. Users have the freedom to choose images that are meaningful to them, increasing the likelihood of password memorability. Additionally, the ability to create passwords by drawing shapes on selected images provides an intuitive and personalized method of authentication. The usability advantages of this approach contribute to improved user experiences and potentially reduce reliance on written passwords or the use of easily guessable alternatives.

Collectively, these papers provide evidence of the benefits associated with graphical passwords. These advantages encompass improved memorability, an enhanced user experience, resilience against specific attack vectors, and the potential for heightened security when compared to traditional alphanumeric passwords. Additionally, they offer valuable insights into the design principles and guidelines for implementing graphical password systems. Even though graphical passwords have been explored and much of the research before, it has never been explored with emergent users in developing countries (i.e. users who are new to technology and may be less educated, among other limitations) to their full potential. In our paper, we have identified a full set of variables that can be experimented with, though we held many of these as control variables for the purpose of the study. In this paper, the aim is to evaluate whether the Passfaces scheme with Unknown faces and Celebrity faces is a more suitable option than PIN's for emergent users which was not covered in the earlier studies.

3 Design

3.1 Study Design

The objective of our study was to compare the performance of emergent users with three types of password interaction techniques, namely PIN, Passfaces with unknown faces, and Celebrities (Passfaces with celebrity faces), where PIN serves as a baseline.

Given that alphanumerical passwords are the current gold standard and the most common type of passwords for many systems, a study could used those as a baseline. However, since we were motivated to work with emergent users, and many emergent users are not familiar with the Latin script, we chose to use PIN as the baseline.

For the purpose of this study, we restricted the **password length** to six digits. In the case of PIN, this gave us a password space of the magnitude 10^6 and a corresponding guessability of $1/10^6$. For Passfaces and Celebrities, after some experimentation, we chose to have a **grid size** of 4×4 (i.e. 16 faces per page), leading to a password space of 16^6 or 1.7×10^7, almost 17 times larger space than the PIN, and a correspondingly lower guessability of $1/(1.7 \times 10^7)$.

Another variable to consider was **repetition of tokens** across digits. By default, for the PIN condition, the same tokens (i.e. the numerals 1, 2, 3 etc.) repeat for each digit and we have no choice (as there are only 10 digits in the decimal number system). But for Passfaces and Celebrities conditions, we had a choice of offering the same (or different) set of faces for each digit. After all, we humans can recognise severally many faces. Offering a different set for each digit removes the possibility of having repeating tokens (the equivalent of a PIN similar to 1111), and we chose to do this. This, we argue improved security of Passfaces and Celebrities condition further.

An interface may display a passwords as typed, or it may **mask** it (either with an asterisk * or with a dot character •). Given that many of our users may be new to password entry (and some new even to text entry), we chose to not mask our passwords.

Lastly, we considered whether to **randomise the layout** of the keypads as suggested by Kirkwood et al. [12]. Such randomisation is known to improve security by reducing the risk of shoulder surfing. However, we were concerned that the performance of the PIN of some emergent users might be seriously affected because of randomisation. Since this was the first study on password usage by emergent users, we chose to not randomise the layout. We note here that we expect randomisation to only affect the PIN condition because numerals have an inherent order in them. Because human faces do not have any inherent order, we feel that it may not affect the Passfaces and Celebrities conditions as much. Randomisation of the layout could be a variable to consider in future studies.

In our study, participants set one password at a time. The participant was first introduced to the study, trained on the first password mechanism and set the first password. After that, the participant was given a distractor task (detailed below), after which, the participant was asked to recall the password. This constituted Recall 1.

Following [18] we chose to use a four-day recall as the primary outcome of our study. After a gap of four days, the participant was invited to recall the same password again (Recall 2), after which the next password mechanism was introduced. We measured whether or not the participant was successful in recalling the passwords if the participant failed, the number of digits that they could not recall, and the time taken by the participant to set the password and to recall it. We provide the details in the Sect. 3.3.

Emergent users may face several challenges in using password mechanisms. Hence they may fail to recall the password that they had set merely after a five-minute distraction task. Hence the success rate in Recall 1 represents the first-time usability of the mechanism.

While some emergent users may fail in Recall 1, given enough training and exposure, they would be able to overcome the issues related to usability. However, in this paper, we are primarily concerned with the memorability of the password mechanisms. Hence we would consider the success rates in Recall 2 as the primary outcome of our study. Further, there are two ways to measure the success rates in Recall 2. If we were to consider the proportion of participants who were successful in Recall 2 compared to all those who participated in the study (i.e. including those who failed in Recall 1), that would represent a cumulative effect of first-time usability and memorability of the password mechanism. On the other hand, if we consider the proportion of participants who were successful in Recall 2 compared to only those who were successful in Recall 1, that, we argue, would represents the 4-day memorability of that password mechanism. Hence, we will treat this as our primary dependant variable.

Thus we conducted a within-subjects, counterbalanced systematic four-day recall study with emergent users, which had one independent variable (the password type) with three levels (PIN, Passfaces and Celebrities), and eight dependent variables, namely Recall 2 success (memorability), Recall 1 success (usability), number of digits recalled correctly in Recall 2, number of digits recalled correctly in Recall 1, time taken to set up the password, time taken for Recall 1, time taken for Recall 2, and number of attempts before a successful recall. Password length, grid size (for Passfaces and Celebrities), repetition of tokens, no masking of passwords, and randomisation of the layout were treated as control variables.

Through this study, we try to explore the following research questions:

- (RQ1) How do emergent users perform in terms of 4-day memorability, first-time usability, and time taken to set and recall passwords three password mechanisms, namely PIN, Passfaces and Celebrities?
- (RQ2) What is the effect of using celebrities (i.e. known, recognisable faces) on Passfaces?

3.2 *Interfaces*

To compare the three interaction techniques, we designed three prototypes, namely PIN, Passfaces with unknown faces, and Celebrities (Passfaces with celebrity faces). Each prototype allows for setting up a password, confirming the password and recalling the password for the given participant. It also logs time for each task, gives input data of the participants (the selected passwords) and the results of each recall. It also logs the number of errors, attempts and wrong entries in each attempt in a tabular

Fig. 2 PIN password design

Fig. 3 Passfaces with unknown faces password design

format. The design of the prototypes was iterated from the insights gathered from a pilot test. For brevity, we present only our final designs below.

The interface for PIN was quite similar to the default keypad layout on smartphones (Fig. 2). As mentioned above, we chose a 4×4 grid for Passfaces (Fig. 3) and Celebrities (Fig. 4) interfaces. To make it easier for the participants, the corpus of photos was categorised for each digit. As can be seen in Fig. 3, in Passfaces, we chose to include old men, old women, young men, young women, old men with turbans, and young girls for digits 1–6. Similarly, as can be seen in Fig. 4, for Celebrities, we chose to include well-known old men, old women, young men, young women, cricketers, and well-known women from other sectors. Each face was carefully selected

Fig. 4 Passfaces with celebrity faces password design

and images that particularly stood out from the rest of the group were discarded. The study was done in the cities of Mumbai and Pune. In case of Celebrities, the selection of images incorporated known faces, including nationally recognisable figures as well locally known faces such as those from Marathi film and television industry. We did a pilot test among our targeted users to ensure that each face was recognisable.

3.3 *Method*

We recruited 66 emergent users with education less than 7 years of schooling, and ensuring an equal proportion of male and female participants. The age range of the participants ranged from 35 to 55 years. Furthermore, we had an equal representation of participants using smartphones and feature phones.

Each participant participated in four sessions (each separated by at the least 4 and at the most 5 days, depending on the availability of the participant) and used the three password mechanisms. The total duration of the study per participant was 12–13 days. The sequence of assignment of the test tasks was counterbalanced across participants to minimize the learning effects. Four identical smartphones were used as hardware to host the test prototypes.

Participants were compensated for their time to attend each session (Rs. 100 per session). In addition, we provided participants with two types of incentives. Firstly, since this was a multi-session study, we provided a "session bonus" to encourage the participant to participate in each session (Rs. 0 for the first session, Rs. 50 for the second session, Rs. 100 for the third session and Rs. 150 for the fourth session). Secondly, we also wanted to encourage participants to remember and recall their passwords correctly. Hence, we provided them a small but important "successful recall bonus" (Rs. 100 if they could successfully recall all 6 digits, Rs. 80 if they could recall 5 digits successfully, and so on). This, we argue, reflects the real world, as something is at stake behind each password. Thus, those who participated in all four scheduled sessions were compensated a minimum of Rs. 700, and if they could recall all the three passwords, a maximum of Rs. 1000.

In the first session, the moderator briefed the participant about the study, explained the compensation scheme and bonuses, and took an informed consent. Participants could choose to opt out of the study at any time if they were not comfortable. Next, to familiarize the participant with the touch interactions, the participant was trained on the first password mechanism that they were going to use. They were also demonstrated to use backspace in the case of any confusion. The training task was not recorded.

Next, the participant was asked to set a password using that mechanism. Right after setting the password, the participant was asked to confirm the password they had chosen.

Commonly a limit of three attempts of password retries is used. Arguably, this is an arbitrary limit and has not been systematically explored, especially with emergent users. Hence we identified the "number of attempts" as a variable of research and did not put an artificial limit. If the participant could not confirm the password, they were given multiple attempts until they could confirm the password correctly. If the participant could not confirm their password at all, we let the participant choose another password using the same mechanism and confirm it. Once the participant had created and confirmed the password successfully, they were given a 5-minute distraction task. The distraction tasks included[1] for Passfaces and Celebrities and[2] where a user is asked to enter the numbers that are appearing on screen in least amount of time for Pins. The distraction tasks are formulated in such a way that it is similar to the type of tasks given as explained above. That is, a distraction task involving faces after setting up graphical passwords and a distraction task using numbers after setting up PIN.

After that, the participant was asked to do Recall 1. Again, the participant was allowed as many attempts as they wanted to recall the password correctly. Only if the participant gave up, this was recorded as a failure. In case of failure, the number of correct digits was recorded. This is the end of the first session.

In the second session, the participant first did Recall 2 of the first password mechanism (again, with as many attempts as desired), set and confirmed a password for the second mechanism, did another distractor task and did Recall 1 of the second mechanism. The third and fourth sessions were similar. After the fourth session, a post-test survey was conducted to understand the user behavior regarding the criteria behind the selection of each type of password. The survey included Likert scales to rate the difficulty of each password mechanism and strategy behind the selection of passwords. Figure 5. summarises the flow of the tasks in the study.

---

[1] games to identify similar faces from a set of tiles: https://www.memozor.com/memory-games/for-adults/people-faces.

[2] number games: https://www.gamesgames.com/game/numbers.

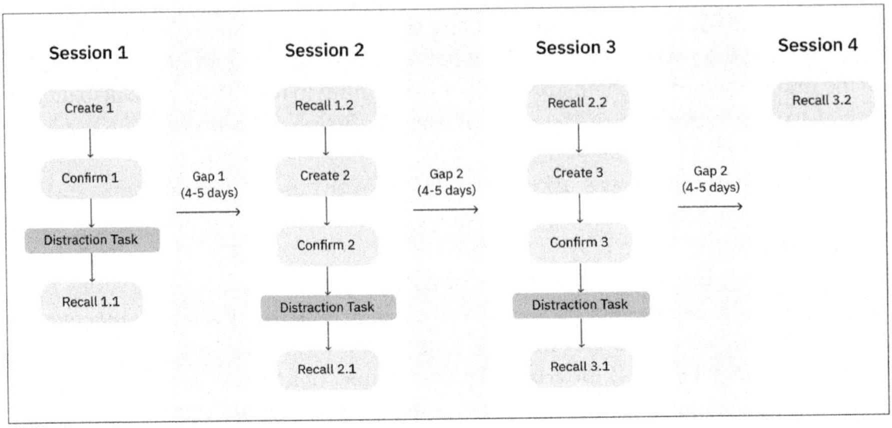

Fig. 5 Overview of study methodology

4 Results

4.1 Success Rates

As discussed above, the success in Recall 1 demonstrates the first-time-usability of the given password mechanism (PIN, Passfaces, and Celebrities), while the overall success rates in Recall 2 demonstrate a combined effect of first-time-usability and 4-day-memorability. If we consider success rates in Recall 2 of only those participants who were successful in Recall 1, we can also isolate the effect of 4-day memorability. It is not surprising that people who failed in recall 1 (immediately after setting the password) will also fail recall 2 (the 4-day recall). Hence, the primary outcome of our study is the percentage of users who were successful in recall 1 and then were also successful in recall 2 after a gap of 4 days. For completeness, though, we report success rates in recall 2 as a percentage. Refer Table 1 for detailed values.

As we can see in Recall 1 that while PIN is the most successful, and Celebrities is the least successful, the difference between the conditions is not significant. 59 users were successful in all three conditions in Recall 1. There is also no significant difference between the conditions in Recall 2 when we consider all users (i.e. the combined effect of first-time usability and 4-day memorability). However, when we considered the Recall 2 success of only those participants who were successful in Recall 1 (which represents the 4-day memorability), we found that Celebrities had the highest 4-day memorability and there is a significant difference ($p = 0.02$) between the conditions. A pair-wise comparison showed that Passfaces was significantly poorer than PIN ($\chi 2 = 3.96$, $p = 0.04$) and Celebrities ($\chi 2 = 5.96$, $p = 0.014$), while there was no significant difference between PIN and Celebrities ($\chi 2 = 0.3$, $p = 0.58$).

We also analysed digit-wise failures as shown in Table 2. We analysed failures in Recall 1 and Recall 2 by coding failed values as ordinal variables. If the user failed

Table 1 Success rates of the three password mechanisms in Recall 1, Recall 2 (if considering all 66 participants) and Recall 2 (if considering only those participants who were successful in Recall 1), which represents the effect of 4-day memorability of that password mechanism

	Recall 1 success First-time usability			Recall 2 success (all 66 participants) First-time usability + 4-day memorability			Recall 2 success (Recall 1 successful participants) 4-day memorability		
	PIN	Passfaces	Celebrities	PIN	Passfaces	Celebrities	PIN	Passfaces	Celebrities
Success	63	62	59	57	48	55	57	48	55
Failure	3	4	7	9	18	11	6	14	4
Total	66	66	66	66	66	66	63	62	59
Proportion (%)	95.45	93.94	89.39	86.36	72.73	83.33	90.48	77.42	93.22
95% CI from	86.95	84.95	79.4	75.85	60.89	72.39	80.40	65.48	83.36
To	98.96	98.06	95.06	92.88	82.07	90.61	95.90	86.16	97.80
	$\chi 2(2) = 1.9984$, $p = 0.3681$			$\chi 2(2) = 4.3638$, $p = 0.1128$			$\chi 2(2) = 7.7011$, $p = 0.0212$		

The last row also presents the results of the $\chi 2$ test

Table 2 Digit-wise failure rates for Recall 1 and Recall 2

	Recall 1 First-time-usability failure analysis			Recall 2 (only users successful in Recall 1) 4-day-memorability failure analysis		
	PIN	Passfaces	Celebrities	PIN	Passfaces	Celebrities
Success	63	62	59	57	48	55
Failures	3	4	7	6	14	4
1 digit failed	0	1	3	0	4	0
2 digits failed	0	0	3	0	7	3
3 digits failed	0	0	1	2	1	0
4 digits failed	0	2	0	1	0	1
5 digits failed	2	1	0	2	1	0
6 digits failed	1	0	0	1	1	0

As we can see, PIN and Passfaces had cases where some users failed to recall 5 and 6 digits. In case of Celebrities, the failure was in fewer digits. While the results are not statistically significant, they indicate that Celebrities may still be aiding recalls

in all 6 digits, we coded it as a value 6. If the user failed in only 1 digit, we gave it a value of 1, and so on. Friedman test was conducted to determine whether there is any difference in digit-wise failure rates of the three conditions. The results show a non-significant difference, $\chi(2) = 0.0083$, $p = 0.995$ in Recall 1 and for Recall 2, $\chi(2) = 2.805$, $p = 0.245$. While the results are not statistically significant, we can observe that in the case of Celebrities, people often forget only one or two digits, and only on 2 occasions have they forgotten more than 2 faces in the whole study. Our study was possibly underpowered to detect this difference, and a larger study could inform us better about this variable.

4.2 *Time Taken*

Some would argue that time taken to set or recall a password is not an important metric in the context of passwords, and the first-time usability, memorability and security are more critical. However, interfaces are often designed to time out if the password is not entered in a specific amount of time. While we had no such time limits, we were mildly surprised by the amount of time our participants needed to set and recall the passwords, and hence we report that data here to inform the design of future password mechanisms for emergent users. Table 3 summarises the time taken to setup, the time taken for Recall 1 and the time taken for Recall 2 for each password condition. Setup time and Recall 1 are reported for all 66 participants in the study. We report Recall 2 times only for the 59 participants who were successful in all three password conditions in Recall 1. In both Recall 1 and Recall 2, the time includes failures (the time after which the participants gave up).

A one-way ANOVA revealed that there was a statistically significant difference in the time taken to set up the three password mechanisms ($F(2,195) = [4.1682]$, $p = 0.01688, \eta 2=0.041$). Post hoc comparisons using Bonferroni correction found that the mean value of time taken was significantly different between PIN and Celebrities ($p = 0.013$, 95% C.I. = [− 146.05, 12.97]). Tukey's HSD Test for multiple comparisons found that the mean value of time taken was significantly different between PIN and Celebrities ($p = 0.012$, 95% C.I. = [− 144.59, 14.43]). The difference in mean setup times between PIN and Passfaces ($p = 0.366$) or between Passfaces and Celebrities ($p = 0.279$) were not statistically significant. In qualitative observations, we noticed that users took more time to choose among their favorite celebrities, which could be the possible reason for the longer time for celebrities. There was no significant difference between the three conditions in Recall 1, though PIN was the fastest.

A one-way ANOVA revealed that there was a statistically significant difference in the time taken for Recall 2 for the three password mechanisms ($F(2,187) = [7.349]$, $p < 0.05, \eta 2=0.07$). Post hoc comparisons using Bonferroni correction found that the time taken for Recall 2 was significantly different between PIN and Passfaces (p = 0.02, 95% C.I. = [− 78.30, 13.47]) and Celebrity and Passfaces ($p = 0.005$, 95% C.I. = [− 74.98, 10.15]). Tukey's HSD Test for multiple comparisons found that the time taken for Recall 2 was significantly different between PIN and Passfaces ($p = 0.02$, 95% C.I. = [− 77.59, 14.18]) and Celebrity and Passfaces ($p = 0.005$, 95% C.I. = [− 74.27, 10.86]). There was no statistically significant difference in mean time taken between PIN and Celebrities ($p = 0.968$).

4.3 *Number of Attempts*

We have identified the "number of attempts" as a research variable. Recall that in this particular study, we did not limit the number of attempts, allowing us to investigate whether emergent users should be given more than the typical three attempts. The findings showed that success was achieved up to the fourth attempt for all three types of passwords. Participants who made more than four attempts could not recall any of the passwords. This also indicates that the current limit of three has to be reconsidered in case of emergent users. We employed Friedman's test to investigate if there were any noteworthy discrepancies in the number of attempts across different password mechanisms. However, no statistically significant differences were observed.

4.4 *Choice of Passwords*

We analysed the passwords entered by the participants qualitatively to identify the presence of pattern-based passwords. We defined a pattern-based password as the one exhibiting obvious visual patterns on the keyboard (like the shape of a pattern lock),

Table 3 Time taken (in seconds) for Setup, Recall 1 and Recall 2)

	Setup Time			Recall 1 Time			Recall 2 Time		
	PIN	Passfaces	Celebrities	PIN	Passfaces	Celebrities	PIN	Passfaces	Celebrities
N	66	66	66	66	66	66	59	59	59
Mean	101.09	138.45	180.06	28.01	45.81	47.56	41.53	87.42	44.85
95%CI from	72.83	104.47	130.38	15.73	34.99	35.8	22.79	63.35	34.66
To	129.34	172.42	230.81	40.28	56.62	59.31	60.26	111.48	55.03
	$(F(2,195) = [4.1682], p = 0.0168$			$(F(2,193) = [3.35], p = 0.06$			$(F(2,187) = [7.349], p < 0.05)$		

Setup time and Recall 1 are for all 66 participants. We report Recall 2 times only for the 59 participants who were successful in Recall 1 for all three password mechanisms. In both Recall 1 and Recall 2, the time includes failures (the time after which the participants gave up repetitive characters/consecutive numbers (in the case of PINs), or the use of the same location inputs (in the case of Passfaces or Celebrities). Two authors of this paper independently analysed each password and classified it as either "exhibiting a pattern" or "not exhibiting any pattern". In case of disagreements, the authors resolved the differences through discussion. The analysis revealed that the percentages of location-based passwords for PINs, Passfaces, and Celebrities were 50%, 33%, and 21%, respectively (N = 66 each). It can be noted that pattern-based selection for PIN is much larger than Passfaces and Celebrities.

Some of the pattern-based passwords in the case of PIN include: (1, 2, 3, 4, 5, 6),(4, 5, 6, 7, 8, 9) which are consecutive numbers, (1,3,4,6,7,9) which are patterns on the screen which is similar to the pattern unlock system, and (3, 3, 5, 5, 9, 9) which show usage of repeated numbers. Considering that pattern-based passwords are less secure against brute force and shoulder surfing attacks [16], we can conclude that Passfaces and Celebrity faces offered higher security compared to PIN passwords on this count.

4.5 *Post-test Survey*

In the post-test survey, several participants reported that their selection of a PIN was influenced by the location of the keyboard, as it facilitated easier recall (but also leads to formation of pattern-based passwords). Fewer individuals reported that they had opted for a randomly generated PIN. For Passfaces, users attempted to identify faces based on distinctive features. If they encountered difficulties (for example, if they could not make up their mind), users resorted to location-based selections, which led to patterns. In the case of Celebrities, there were very few instances of location-based selections. Participants predominantly chose celebrities that they knew or could identify, and based on personal preferences. Qualitatively, we observed that the participants exhibited a high level of enthusiasm and eagerness to try out the new password systems, and particularly with Celebrities.

4.6 *Popularity of Selection*

Table 4 lists the number of times a particular face was chosen by participants in the Celebrities condition (out of 66) (please also refer to Fig. 4). This revealed the popularity of 12 personalities who were chosen by 10 or more participants, namely Amitabh Bachchan (28), Salman Khan(27), Narendra Modi (21), Sindhutai Sakpal (20), Sai Tamhankar (14), Aishwarya Rai (14), Virat Kohli (14), Sachin Tendulkar (13), Lata Mangeshkar (12), Harbhajan Singh (12), Indhira Gandhi (10), Sachin Pilgaonkar (10). As many Indians will readily agree, while some of these individuals can be recognised all over the country, some will perhaps be recognised only within Maharashtra (or perhaps only within the cities of Mumbai and Pune). These choices suggest that localisation works and the corpora of Celebrity images could be extensively localised. Such localisation could enhance security further, limiting the number of users a potential hacker can target.

We can also observe that the participants' choices were also influenced by the relative popularity of the celebrities within that specific page. For instance, on Page 1, Amitabh Bachchan and Narendra Modi were both very popular compared to other celebrities on that page and together account for over 74% of the choices by our participants. While using popular personalities is what makes the Celebrities password mechanism easier to use, when their popularity is so high, it is also a potential threat to security. This suggest that popularity of a personality is not the only criteria for including a celebrity in a password mechanism. Instead, an attempt should be made to design a more "balanced" page, where the popularity across personalities on a particular page is a little less skewed. A designer may thus be required to not select some of the most popular personalities.

It is worth noting that the position on the page also influences the choice. The total number of selections in the first four positions, i.e. the first row of all pages, account for about 48% of the choices (where it should have been closer to 25%). This indicates that there is a positional bias in the choice. Whether this is coincidental (as the most popular celebrity choices were also in the first row), or more systematic needs to be investigated in future research.

We tried to analyse if the popularity of a celebrity also makes them more memorable at the time of recall. We grouped choices into three groups—those that were chosen 11 or more times, those who were chosen 6–10 times, and those who were chosen between 1 and 5 times. We also analysed the error rates for those faces in their last recorded attempt at recall. For participants who succeeded in Recall 1, we considered last attempt in Recall 2. Those who failed in Recall 1, we considered their last attempt in Recall 1. We found the error rates to be 12%, 12%, and 2.75% respectively for the three groups (Fig. 6). As can be expected, the popular celebrities were also memorable. This is also evident from the popularity versus memorability scatter plot show in Fig. 7a, in which each dot represents one face. The x-axis represents the number of participants who picked that face. The y-axis represents the memorability of that particular face. Less popular faces are somewhat less memorable, though the effect sizes are quite small (as indicated by the R^2 values).

We also performed a similar analysis of the faces presented in the Passfaces condition (Table 5, read together with Fig. 3). As we can see, only 5 faces from this group were chosen by 10 or more participants, indicating a broader coverage. Like in case of Celebrities, the position on the page also influences the choice in Passfaces too. The total number of selections in the first four positions, i.e. the first row of all pages, account for about 45% of the choices (where it should have been closer to 25%).

Table 4 Popularity among the celebrities

Position in page	Page 1	Page 2	Page 3	Page 4	Page 5	Page 6	Total
Position 1	28 (1)	7	3	3	3	8	49
Position 2	21 (3)	8	10 (12)	6	14 (7)	2	61
Position 3	2	12 (9)	27 (2)	0	5	0	46
Position 4	1	5	1	14 (5)	6	7	34
Position 5	0	2	5	5	0	2	14
Position 6	0	0	7	3	0	1	11
Position 7	1	1	2	1	0	1	6
Position 8	1	2	0	2	12 (10)	1	18
Position 9	2	1	1	0	1	4	9
Position 10	1	3	0	14 (6)	5	20 (4)	33
Position 11	1	4	5	0	3	2	15
Position 12	1	3	1	3	0	2	10
Position 13	4	2	1	5	3	9	24
Position 14	0	3	3	0	13 (8)	3	22
Position 15	0	2	0	2	1	1	6
Position 16	3	10 (11)	0	8	0	3	24

The table shows the number of participants (out of 66) who chose that particular celebrity. We have ranked the most popular 12 names in the list. Their names are: (1) Amitabh Bachchan, (2) Salman Khan, (3) Narendra Modi, (4) Sindhutai Sakpal, (5) Sai Tamhankar, (6) Aishwarya Rai, (7) Virat Kohli, (8) Sachin Tendulkar, (9) Lata Mangeshkar, (10) Harbhajan Singh, (11) Indira Gandhi, (12) Sachin Pilgaonkar

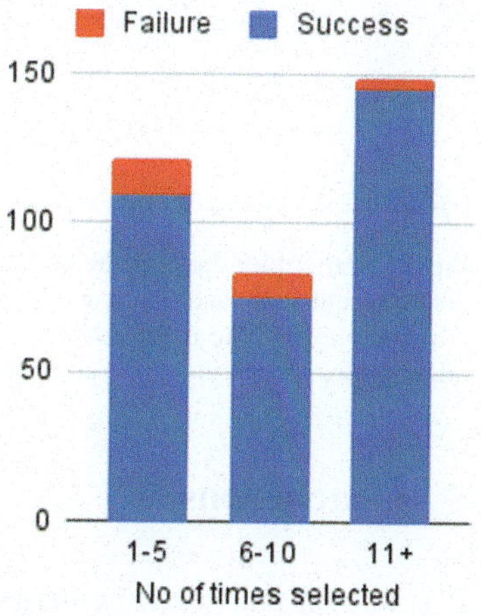

Fig. 6 Comparing success and failures of the last attempts of recall where celebrity faces are categorised by popularity. Failure rates were 12% for faces that were picked 1–5 times and 6–10 times. But for faces that were picked 11 times or more, the failure rates were 2.75%

Table 5 Popularity of the unknown faces among the Passfaces

Position in Page	Page 1	Page 2	Page 3	Page 4	Page 5	Page 6	Total
Position 1	11(4)	6	8(8)	7	3	12	47
Position 2	9(6)	5	4	5	6	17(2)	46
Position 3	4	2	6	3	19(1)	3	37
Position 4	8(8)	13(3)	5	11(4)	5	7	49
Position 5	1	2	2	1	6	2	14
Position 6	4	3	5	5	1	2	20
Position 7	1	1	2	1	0	1	5
Position 8	3	9(6)	3	2	0	2	19
Position 9	2	1	2	4	3	7	19
Position 10	0	2	2	1	5	5	15
Position 11	6	1	1	2	2	2	14
Position 12	4	2	5	2	2	1	16
Position 13	1	1	3	7	2	1	15
Position 14	5	6	7	8(8)	3	1	30
Position 15	2	6	2	0	2	2	14
Position 16	3	4	0	6	0	1	14

Brackets indicate ranks

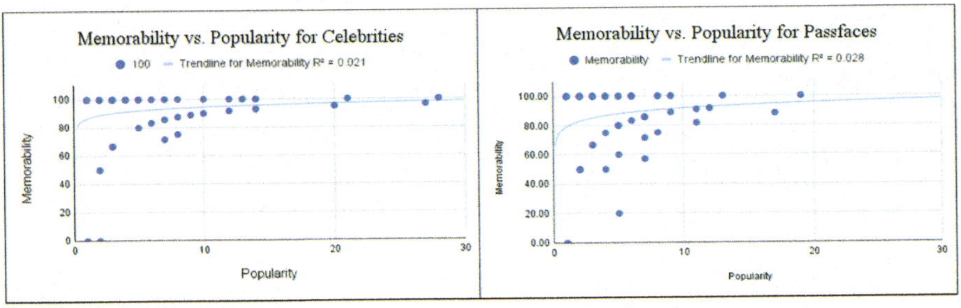

Fig. 7 Scatter plot showing the popularity versus memorability for Celebrities and Passfaces. Each dot represents one face. The x-axis represents the number of participants who picked that face. The y-axis represents the memorability of that particular face. Less popular faces are somewhat less memorable, though the effect sizes are quite small (as indicated by the R^2 values)

5 Conclusions

It is known that several emergent users struggle with text entry and related tasks. They struggle even more in account-holding tasks such as logging in with passwords [7]. We can address these challenges by designing more suitable what-you-know password mechanisms.

We conducted a systematic, lab-based, within-subjects, counterbalanced study with emergent users to compare PIN-based passwords with Passfaces (with unknown faces) and Celebrities (Passfaces with celebrity faces). By design, we made sure that the password space of Passfaces and Celebrities was larger than PIN. While the tokens were repeated in case of PIN, they were different for each digit of the password in case of Passfaces and Celebrities. This enhanced their security further. We found that though the first-time usability was comparable between conditions, Celebrities, and PIN were significantly more memorable than the traditional Passfaces. While Celebrities had longer setup times, this was because users spent time choosing their passwords carefully. Celebrities and PIN also were significantly faster than Passfaces during Recall 2. We found that the Celebrities mechanism is less prone to pattern-based passwords than PIN or Passfaces, though participants were still to favour the faces in the first row.

It seems to be the case that the Celebrities mechanism presented in this paper is at least as usable and memorable as the numerical PIN. In addition, it has other advantages. It already has a larger password space, and it has the potential for being scaled up further. The choices of celebrities need to be made carefully, making sure that many of the chosen individuals are widely recognizable. Locally known faces are also likely to be picked up, opening opportunities for localisation (and thereby improving security). We found that there was a weak positive correlation between the popularity of celebrities and their memorability—the more popular celebrities were somewhat more memorable (as were the unknown faces). However, including very popular celebrities could be counterproductive with respect to security as that would make the password more guessable. The whole page of faces needs to be balanced out in terms of comparable popularity of the individuals presented.

We expect that Celebrities can be made more secure by randomizing its layouts. This will not only help reduce shoulder surfing attacks, but it will also help reduce the "first-row" bias, and other pattern-based approaches for password selection. To what extent these choices will affect usability and memorability needs to be systematically investigated in further research.

Though several of our users used a smartphone, only a small number used a locking mechanism. These users displayed limited awareness of potential security threats associated with their devices. To instill a sense of challenge and help the participants maintain a serious approach towards the study, we provided two small monetary incentives—one for participation throughout the study and one for remembering the passwords they had set up. A noteworthy observation from the study was the remarkably low dropout rate throughout the entire duration. Most participants attended all four days of the study, with only a small number who were unable to continue due to personal commitments.

Several participants employed intriguing strategies when selecting their passwords. Some users developed unique associations by focusing on specific characteristics, such as choosing individuals wearing certain-colored clothing or displaying particular facial expressions, as a technique to aid recall. These findings highlight the participants' creativity and adaptability in leveraging visual cues to enhance their password memorability.

Several limitations should be acknowledged in this study. Firstly, due to time constraints, the number of recalls and the duration between Recall 1 (immediate recall) and Recall 2 (after 4 days) were limited. Including an additional recall, after a much longer duration could have provided greater insights into memorability. Therefore, the results pertaining to long-term memorability should be interpreted with caution. Secondly, the study focused solely on participants from just two cities, limiting the generalisability of the findings to other regions. Including participants from a wider geographic range would have provided a more comprehensive understanding of how the localization of the corpus influences password selection and memorability. Furthermore, it was observed that certain celebrities were chosen by a significant number of participants, potentially influencing the overall distribution of password selections. Conducting a follow-up study that excludes these highly popular celebrities would yield a more stable and unbiased distribution of password choices. These limitations highlight the need for future research to address these constraints by including larger sample size, incorporating additional recall sessions, diversifying participant demographics, and exploring the impact of excluding influential celebrities.

References

1. Andriotis P, Oikonomou G, Mylonas A, Tryfonas T (2016) A study on usability and security features of the android pattern lock screen. Inform Comput Secur 24(1):53–72
2. Bishop M, Klein DV (1995) Improving system security via proactive password checking. Comput Secur 14(1995):233–249
3. Brostoff S, Angela Sasse M (2000) Are Passfaces more usable than passwords? A field trial investigation. In: People and computers XIV-usability or else! Proceedings of HCI 2000. Springer, Berlin, pp 405–424
4. Chiasson S, Van Oorschot PC, Biddle R (2007) Graphical password authentication using cued click points. In: ESORICS, vol. 7. Springer, Berlin, pp 359–374
5. Darren D, Fabian M, Reiter Michael K et al (2004) On user choice in graphical password schemes. In: USENIX security symposium, vol 13, p 11
6. De Angeli A, Coventry L, Johnson G, Coutts M (2003) Usability and user authentication: pictorial passwords vs. PIN. Contemp Ergon 2003:253–258
7. Devanuj, Joshi A (2013) Technology adoption by 'emergent' users: the user-usage model. In: Proceedings of the 11th Asia Pacific conference on computer human interaction, pp 28–38
8. Dunphy, P, Yan J, McBurney P (2008) The emperor's new security mechanism: a graphical password generator. In: Proceedings of the 2008 workshop on New Security Paradigms (NSPW). ACM, pp 57–66
9. Farik M, Lal N, Prasad S (2016) A review of authentication methods. Int J Sci Technol Res 5:246–249
10. Gokhale A, Waghmare V (2016) The shoulder surfing resistant graphical password authentication technique. Procedia Comput Sci 79(2016):490–498
11. Jermyn I, Mayer A, Monrose F, Reiter MK, Rubin A (1999) The design and analysis of graphical passwords. In: 8th USENIX security symposium (USENIX Security 99)
12. Kirkwood D, Tombul C, Firth C, MacDonald F, Priftis K, Mathis F, Khamis M, Marky K (2022) PIN scrambler: assessing the impact of randomized layouts on the usability and security of PINs. In: Proceedings of the 21st international conference on mobile and ubiquitous multimedia, pp 83–88
13. Markert P, Bailey D, Golla M, Dürmuth M, Aviv A (2021) On the Security of Smartphone Unlock PINs. ACM Trans Privacy Secur 24(11), 1–36, https://doi.org/10.1145/3473040
14. Meng Y (2012) Designing click-draw based graphical password scheme for better authentication. In: 2012 IEEE Seventh international conference on networking, architecture, and storage. IEEE, pp 39–48

15. Morris R, Thompson K (1979) Password security: a case history. Commun ACM 22(11):594–597
16. Shammee TI, Akter T, Mou M, Chowdhury F, Ferdous MS (2020) A systematic literature review of graphical password schemes. J Comput Sci Eng 14(4):163–185
17. Thorpe J, van Oorschot PC (2015) An empirical investigation of user choice in graphical passwords. Proceedings of the 2015 workshop on New Security Paradigms (NSPW). ACM, pp 47–58
18. Tupsamudre H, Dixit A, Banahatti V, Lodha S (2017) Pass-roll and pass-scroll: new graphical user interfaces for improving text passwords
19. Wiedenbeck S, Waters J, Birget J-C, Brodskiy A, Memon N (2005) PassPoints: design and longitudinal evaluation of a graphical password system. Int J Human-Comput Stud 63(1–2):102–127

Is JOMO the Answer to FOMO? Collaborative Digital Detox from Social Media for Adolescents

Anjali Mishra, Sanchita Jagdishkumar Bagre, Wricha Mishra, and Sumita Sharma

Abstract Adolescent's usage of social media impacts their daily lives—from procrastination to FOMO (Fear of Missing Out), and other forms of peer pressure. While digital detoxing is a popular solution, it can, however, be an isolating experience. In this paper, we ask—Can JOMO (Joy of Missing Out) be the answer? That is, can we motivate adolescents towards digital detoxification, in a collaborative way? Following a Double Diamond design process, we conducted a user study with adolescents (15–17 years old) in India investigating their social media experiences and challenges. Our findings highlighted the need to combine tangible rewards and offline activities with digital detox; based on which we designed an application, called JOMO, to promote a balance between social media usage and real-life interactions through collaborative digital knitting. Through our work, we highlight the need for innovative solutions to address the negative impacts of social media and FOMO on adolescents and recommend collaborative approaches toward digital detoxing.

Keywords Adolescents and social media · Tangible detox · Time management application

CCS Concepts • Human-centered computing • Human computer interaction (HCI) • HCI design and evaluation methods • User studies

---

A. Mishra · S. J. Bagre · W. Mishra (✉)
User Experience Department, Institute of Design, MIT ADT University, Pune, India
e-mail: m.wricha84@gmail.com

A. Mishra
e-mail: mishra31anjali@gmail.com

S. J. Bagre
e-mail: sanchitabagre@gmail.com

S. Sharma
University of Oulu, Oulu, Finland
e-mail: sumita.sharma@oulu.fi

A. Joshi and G. R. Sim (eds.), *Proceedings of the 14th Indian Conference on Human-Computer Interaction*, Studies in Computational Intelligence 1162,
https://doi.org/10.1007/978-981-97-4335-3_5

1 Introduction

In India, approximately 95% of adolescents demonstrate extensive engagement with social media platforms, maintaining active online profiles [33]. A study conducted by the Pew Internet and American Life Project indicates a significant rise in content creation among adolescents aged 12–17 years engaged in online activities, with 64% actively generating content, a notable increase from 57% in the 2004 survey [22]. This escalating engagement with social media has prompted concerns. The ramifications extend beyond mere time consumption; they encompass significant issues such as sleep deprivation, heightened anxiety, and depression, all of which may subsequently impact academic performance and interpersonal relationships [16]. Various factors, including family background, educational environment, and mental health, contribute to the susceptibility to social media addiction [39]. There is an imperative to research the prevalence of social media addiction among this demographic and to provide them with strategies for prevention and management [35].

In 2004, McGinnis introduced the term Fear of Missing Out (FOMO), which is defined as the constant worry that others are enjoying experiences without us [13]. FOMO can take two forms: feeling left out of an opportunity or developing compulsive behaviors to fit in [8]. The rise of social media has extended FOMO into the virtual world. Studies have reported that adolescents are hooked to Social Networking Sites (SNS) such as Instagram, WhatsApp, and YouTube for escape from daily routines and entertainment purposes [2, 19, 39]. In Fact, they've become essential for meeting their social needs. Furthermore, constant exposure to selective portrayals of other's lives on social media platforms intensifies feelings of exclusion and fuels FOMO [15, 28]. Active participation such as posting and sharing content, has the potential to alleviate FOMO [5, 14]. This again results in an increase in social media usage time, thus increasing the possibility of technology addiction, a commonly found phenomenon in adolescents [17]. Both FOMO and problematic internet usage have a noteworthy negative impact on individuals' mental health [26]. A significant relationship between psychological distress and social media addiction among adolescents is reported [21]. Adolescents being the prime age for developing (harmful) coping mechanisms and getting influenced, become an easy target [5]. Thus, it is important to educate them and to provide strategies to deal with such situations.

Many solutions and strategies are proposed in order to help users from the negative effects of social media usage. Digital detox is one such phenomenon. People today are recognizing how all-consuming social media engagement can be; they prefer to limit or restrict their time online. A "digital detox" refers to the intentional reduction or cessation of electronic device usage, aiming to mitigate potential over-reliance on or overexposure to technology [5]. Software tools designed to assist in this endeavor are termed "digital detox applications."

In the milieu of the FOMO phenomenon, the concept of the Joy of Missing Out (JOMO), originally attributed to Anil Dash circa 2012, has surfaced as a noteworthy construct [34]. In the same article, Christina Crook's 31-day social media detox confirmed that embracing JOMO heightens joy and fosters meaningful connections. Her experience emphasizes the importance of deliberate digital choices, warning against mindlessly following FOMO-driven trends. JOMO operates on a similar idea: When people are aware of what truly makes them happy, they can confidently pursue that path without feeling guilty or doubtful about their ability to find joy for themselves [18]. When an individual establishes a clear priority list, they cease to experience the compulsion to be present everywhere and, instead, give precedence to their personal comfort and preferences. This proactive approach aids in mitigating the sensation of FOMO [26]. Further, addressing FOMO can lead to a reduction in compulsive checking of social media updates [1]. This, in turn, promotes a more balanced and intentional use of digital platforms. If adolescents receive guidance on effectively allocating their time between online and offline activities, they have the potential to cultivate a valuable habit of balancing digital and real-life interactions early in life. The instillation of these principles during youth can serve as a motivating force, propelling the younger generation toward comprehensive personal development [10].

Social media deeply influences various facets of adolescents' lives, from academics and family relations to mental health [17]. In our post-pandemic landscape, its accessibility and influence have grown even more pronounced [28]. In this paper, we present our work on exploring healthier alternatives for social media engagement and incorporate them with a digital detox. Our study employs a mixed-methods approach, integrating the Double Diamond framework and systematically progressing through the four phases: Discover, Define, Develop, and Deliver. The outcome of our work is the JOMO application- intended to equip adolescents with strategies to balance real-life activities, improve focus, and promote present-moment awareness, fostering a healthier approach to digital interactions. This research sets the groundwork for future investigations into strategies that bolster healthier digital habits and overall mental well-being so that we all can feel the joy of missing out over the fear of missing out.

2 Related Works

Studies have established a significant correlation between adolescents' excessive use of social media platforms, such as Facebook and Instagram, and the development of symptoms associated with anxiety and depression [6, 19, 31]. In this section, we first present the previous work on adolescents and social media, and then Human–Computer Interaction (HCI) research on digital detoxes.

2.1 Adolescents' and Social Media

Research indicates that adolescents use screens for an average of 7.5 h daily. This duration amplifies during weekends, often increasing by 1–3 h and occasionally exceeding 5 h per day [5, 32]. Adolescents seek inclusivity, desiring updates on their peers and environment. Engaging via shared posts, likes, or mere scrolling offers insights into others' lives, fostering a sense of connection not just with friends but also with celebrities and influencers [8, 19]. Such constant immersion in social media during early adolescence can influence the brain's perception of social rewards and repercussions [18]. A prevalent manifestation of this excessive engagement is "phubbing," where individuals overlook in-person interactions in favor of their devices. Potential mitigation strategies include restricting mobile usage, increasing awareness, introducing school-based programs, and promoting public service announcements [20].

Smartphones have secured a pivotal role in the lives of adolescents, cultivating a pervasive sense of attachment [37]. In one study, 34 adolescents who participated in a six-month smartphone-free sailing trip, when assessed for fear of missing out, reported experiencing heightened anxiety within three weeks of separation from their devices [38]. Research illuminate's adolescents' engagement with social media, characterizing dynamic roles as 'viewers' and 'contributors.' 'Viewer' activities encompass actions like receiving likes and scrolling but engender prolonged feelings of envy and inadequacy. While 'contributor' behaviors include the use of filters and content curation, temporarily boosting self-esteem and social popularity [4]. Furthermore, an investigation involving 180 participants (ages 18–40 years), assessed through instruments such as SAS-SV, PSQI, and PWB, unveiled a positive and significant correlation between smartphone addiction and diminished sleep quality, along with a significant and adverse correlation between smartphone addiction and mental health [17], In the same investigation, Smartphone addiction was identified as an independent and significant predictor of hypertension in adolescents' well-being (aged 12–15 years). Adolescents experiencing anxiety often employ frequent social media use as an active coping mechanism [9].

In promoting Adolescent mental health, it is imperative to motivate a balance between technology usage that augments the overall quality of life and the capacity to function effectively offline [6]. Additionally, the adoption of short-term voluntary disconnection emerges as a viable strategy to address excessive smartphone engagement [18]. In sum, these findings underscore the necessity of monitoring social media consumption, provision of mental health support, and cultivation of responsible technology usage habits among adolescents.

2.2 *HCI in Digital Detox*

The role of social networks in shaping student behavior is noteworthy, accounting for a significant 26.4% [20]. Research in HCI has explored various solutions and strategies to help users from the negative impacts of social media use. Digital detox is one such phenomenon where there is a periodic disconnection from technology, that is, a hiatus from electronic devices [6, 10]. Many teenagers are unfamiliar with the concept of digital detox but are eager to try it as a way to limit their social media consumption. This includes expanding their *no-phone areas* at home and dedicating one day per weekend as a phone-free day [36]. A digital detox does not mean completely avoiding one's phone but using it mindfully to maintain a balance between one's real life and virtual life [9, 25]. It should be viewed as a valuable concept in technostress research, offering a preventive and strategic element to address technostress at both individual and organizational levels.

Numerous digital detox applications exist, addressing mobile phone addiction among users, particularly adolescents, who employ these apps for effective time management and to encourage occasional breaks from their devices to enjoy real-life experiences [25, 32]. While these apps are useful in reducing screen time on phones, they can also become a boring routine after a certain time due to a lack of user engagement. Thus, highlighting the need for a study on existing digital detox applications used by adolescents to reveal their unmet expectations. Among various factors associated with problematic social media use, it is reported that FOMO is one of the major factors [24]. Encouraging mindfulness of the present can shift focus from distractions. "JOMO," remains underexplored in research. Transitioning from "FOMO" to "JOMO" amid the pandemic could be beneficial [14].

Our work caters specifically to adolescents and their need for social interactions yet also for digital detoxing—both in tangible and online aspects. Previous research provides evidence that interactions that include achievement and progression elements satisfy users' needs for competence, autonomy, and relatedness [7], thus, we incorporate these elements in our explorative work, as described next.

3 Research Methodology

Our research adopted a systematic approach where outcomes from one phase seamlessly informed the subsequent steps. We incorporated the Double Diamond framework which encompasses four distinct phases: Discover, Define, Develop, and Deliver. Figure 1 provides a visual representation of research methodology.

Starting with a qualitative paradigm, the Discover Phase is dedicated to an exhaustive exploration of the social media landscape. In Sects. 2.1 and 2.2, we provide a comprehensive review of extant research on social media with adolescents, focusing on the mental health implications and considerations regarding digital detoxification. We conducted qualitative interviews with 33 participants (19 females, 14 males) aged

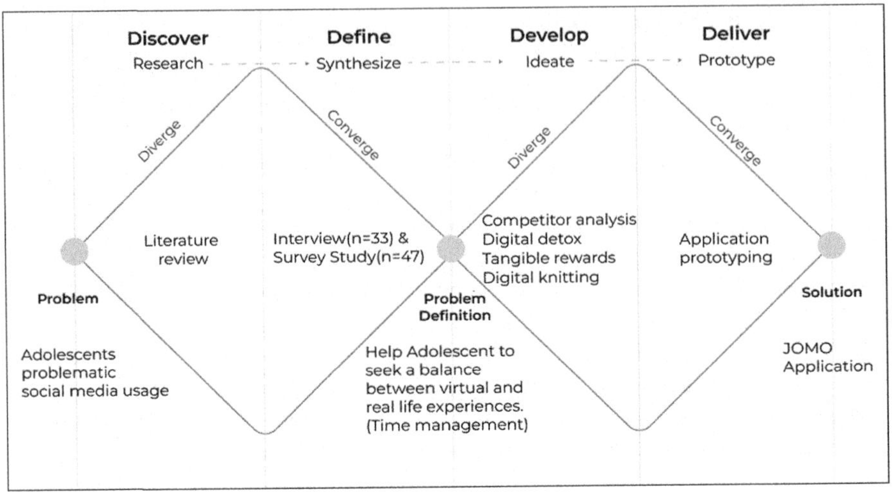

Fig. 1 Research Methodology

13–19 years from Pune, India; recruited via snowball sampling method. Our structured questionnaire probed their engagements with social media, time management practices, and the overarching influence of SNS. Key insights emerged during the interviews. Notably, about 70% of adolescents' secure personal devices between ages 15–17 years, with the younger ones typically accessing their parents' devices. This finding steered our focus toward 15–17-year-olds as the primary user group for subsequent phases. We observed a predilection among participants for platforms like WhatsApp, Instagram, YouTube, and Spotify. Concurrently, apps like Forest, Digital Detox: Focus and Live, Habitica, and Digital Wellbeing emerged as preferred tools for managing screen time. We analyzed social media use, feelings of FOMO, challenges with and strategies for time management, and the impact on lifestyles, using thematic analysis of extensive qualitative data. The interview findings provided insights into why and how adolescents use social media. In the next phase, we conducted a statistical survey to quantify these patterns.

In the Define phase, we undertook a quantitative approach to understand the statistics of social media usage: we conducted an online survey targeting 15–17-year-olds via snowball sampling method in Pune, focusing on those who use WhatsApp, Instagram, YouTube, and Spotify, as they were reported to be the most used apps in the interviews. Excluding those without their own smartphone access, a total of 47 participants (33 females, 15 males) from Fergusson College and MIT ADT University answered the survey. The survey questions were designed based on interview findings, covering demographics, social media experience, and mental health related to phone usage. We analyzed the data for the frequency of responses to closed-ended questions. Key insights that emerged during the survey highlighted the statistics of social media consumption and its impact on adolescents and teenagers.

In the Develop phase, four digital detox and time management apps (Forest, Digital Detox: Focus and Live, Habitica, and Digital Wellbeing) were analyzed. These apps were selected based on the preferences expressed by adolescents during the interviews. A competitive analysis of the apps was conducted, exploring their features, limitations, and usage. Our goal was to analyze the features and limitations of these applications and employ the findings in designing a solution that promotes the 'Joy of Missing Out' (JOMO). A week-long analysis was undertaken by two researchers, documenting primary and secondary features, limitations, usage factors, and user impact (Table 2). Gaps were identified to guide an improved design solution.

The culmination of our research endeavor was the Deliver phase, the data from interviews, surveys, and competitive analysis inspired the creation of 'JOMO,' an app to help adolescents manage their time and detox from phone usage. Taking inspiration from the Forest App theme, after ideation we selected digital knitting as our representation theme to virtually inform users about time constraints and to make the approach of digital detox interesting and easy to understand. 'JOMO' incentivizes digital detox with rewards for user retention, promoting outdoor activities and social interactions. It features a reward system based on screen-free time, redeemable for event coupons, and supports collaborative detox with friends. Section 4.4 provides an in-depth exploration of the 'JOMO' app, its features, and functionality.

4 Findings

In this section, we present the findings of our exploratory study, and in Sect. 4.4, we present the outcomes-, that is, the application design in the Deliver phase.

4.1 Findings from the Discover Phase

The findings from the interviews showcase several noteworthy insights into the everyday online experiences of adolescents. Almost half of our participants (48%) acknowledged social media's negative impact on their mental well-being and mentioned having experienced FOMO. As one participant exclaimed, "*I many times feel that I want to restrict social media usage time, but I find myself going back to same platforms after 5 min again.*" (P1), and another mentioned, "*Many times I find myself overthinking due to something I saw my friend doing on social media, it gets too competitive*" (P2). Participants also reported feelings of loneliness and anxiety arising from their social media interactions, as stated by P3, "*Social media has made me feel lonelier more than anyone else*". Furthermore, a notable 53.4% of the participants admitted to engaging in various activities, such as following recommended content or conforming to trends encompassing fashion, social gatherings, online courses, dancing, and singing, to fit into social media groups. For instance, participants mentioned, "*I sometimes plan an outing with my friends just so that I*

Table 1 Interview and survey findings

Objective of social media engagement (Sect. 4.1—Discover Phase)	Statistics for objectives (Sect. 4.2—Define Phase) (%)
Fashion inspiration	53.8
Explore dining venues	87.3
Discover travel destinations	73.3
Media content recommendations	68.7
Staying informed about upcoming events	36.4

can visit a place I saw on my Social media" (P4), and "*Any series that is hyped up on social media creates an urge in me to see it, I want to know what and why they are discussing about that show or movie.*" (P5). This behavior can be seen as a manifestation of the second form of FOMO, characterized by pretentious actions aimed at fitting in, as introduced in Sect. 1 of this paper.

Regarding the practice of "phubbing," a substantial 64.3% of respondents reported unintentional instances of this behavior while using their devices, which had detrimental effects on their family and in other social interactions. As revealed by P6, "*My parents have this mandatory complaint that I ignore them and keep using my phone at home and in social gatherings.*" and P7 "*I had a fight with my boyfriend, because even though we met after 2 long weeks, he was more interested in his phone than our conversion*". We found that when seeking recommendations or guidance via social media, a significant portion of individuals turned to influencers for a range of purposes, including seeking fashion inspiration, exploring dining venues, discovering new places to visit, media content recommendations, and staying informed about upcoming events (as shown in Table 1). Participants also mentioned developing several negative coping mechanisms in themselves, such as pretentious behaviour—"*Funny how Social media was meant to connect us but now it makes one feel out of place, at least to me.*" (P8), and issues related to addiction, "*During exams I often deactivate my social media account so as to focus on studies, it does help but only for a while soon something distracts and you get this greed of checking social media for once, it never restricts to once*" (P9). Thus, while social media aids in forming friendships and staying updated, it also negatively affects adolescents' mental health and well-being. Our findings from the interviews highlight the need for a comprehensive understanding of this complex relationship.

4.2 Findings from the Define Phase

In the define phase, we conducted a survey with 47 adolescents (33 females and 15 males, ages 15–17 years). Survey results offer significant insights into the social media habits of the participating teenagers, with 70.2% indicating they check their phones 2 times in every five minutes, or 24 times in an hour. Another noteworthy

Table 2 Competitor analysis

Application	Forest	Digital detox: focus and live	Habitica	Digital wellbeing
Problems addresses in application	To help people with phone addiction and focus problems	Lack of motivation to do real life task without phone engagement	To do real life tasks in a gamified way	Acts as a third-party app
Strengths	It plants trees in real life	Shows motivational messages while on focus mode	Makes tasks virtually rewarding in a game	Works in background
Weakness	In—app purchases	Points earned have no rewards Have to give real money as penalty	Confusing UI	Only insights on phone usage, no method to help
Primary features				
Gamified focus mode	Yes	Yes	Yes	No
Live Tasks to focus on	No	Yes	Yes	No
Motivation for hobbies	No	Yes	No	No
Rewards	Yes	No	No	No
Motivation to stick to timer set	No	Yes	No	No
Background Music to focus on	Yes	No	No	No
Temporarily close application	Yes	Yes	No	Yes
Secondary features				
Invite Friends	Yes	No	No	No
Activity Chart	Yes	Yes	No	Yes
Bedtime Mode	No	Yes	No	Yes
Do not Disturb mode	No	No	No	Yes
Apps selection for not using	Yes	Yes	No	Yes

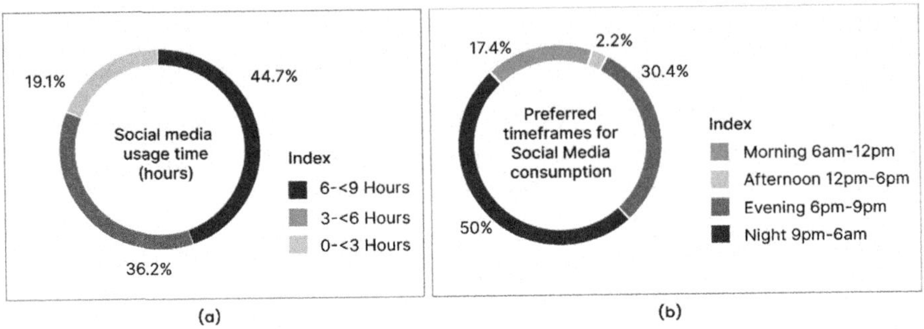

Fig. 2 **a** Social media usage time and **b** preferred time frame for social media consumption

discovery is that 44.7% of survey participants dedicate approximately six to nine hours daily to social media, primarily during the hours of 8 p.m. to 1 a.m., as depicted in Fig. 2. Additionally, 48.9% maintain multiple active social media accounts, which could contribute to an increase in time spent on social networking sites. Furthermore, 67% of participants reported experiencing FOMO (Fear of Missing Out), and 21.3% occasionally felt FOMO after seeing friends' social media posts about an activity.

Among those surveyed, Instagram emerged as the top preference for social media, with 78.7% favoring it, followed by YouTube (44.7%), Snapchat (29.8%), and Spotify (2.1%). As participants mentioned in Sect. 4.1 section purpose of social media usage, in this phase we collected statistics on the same area. We found that when seeking social media, Adolescent's majorly root for exploring dining venues (87.3%), followed by recommendations on new places to visit (73.3%), media content recommendations (68.7%), fashion ideas (53.8%), and staying updated on events (36.4%), as shown in Table 1. Each Adolescent's followed at least three prominent influencers and 58.9% accepted seeking their advice.

However, as mentioned in the interviews, there are also negative influences rampant in social media use and experience. Survey participants noted engaging in pretentious behavior (36.4%), experiencing depression (21.7%), and 31.9% of them mentioned facing addiction issues, including alcohol and substance abuse. This data underscores the importance of promoting a balanced social media approach. Consequently, we advocate for digital detoxing among adolescents, to harmonize real-life activities with digital engagement. In this vein, we conducted a competitive analysis of currently available digital detoxing applications, focusing on those mentioned by the interview participants.

4.3 Findings from the Develop Phase

Post-survey, we delved into existing digital detox applications through a competitive analysis of four apps,—Forest, Digital Detox: Focus and Live, Habitica, and Digital Wellbeing, selected based on their recognition among interviewees. An assessment of their features, detailed in Table 2, gave insights into their usability, innovation, and distinctiveness.

The theme of the Forest application is centered around "growing" a virtual plant. Forest distinguishes itself with a unique offline reward system. Users "plant" a virtual tree for focused work hours and earn rewards by accumulating virtual coins, which can then be used to make real-world donations to plant trees. However, these intangible rewards don't significantly boost user retention, and the application is restricted to an intangible form of gratification. This key finding motivated us to consider strategies to incorporate tangible rewards in our application design (mentioned in Sect. 4.4).

Digital Detox: Focus and Live employs motivational notifications as its main strategy, but uses a penalty-based system where users have to pay actual money if they miss their targets. Habitica encourages users to gamify their everyday tasks with virtual rewards for task completion. Digital Wellbeing application is more of a background app that provides insights into phone usage but no actual strategies for detoxing. Except for Habitica, all others allow blocking distracting apps on one's phone. Other shared features across all four applications include activity charts, app usage statistics, bedtime modes, and Do Not Disturb settings. Each app provides its own unique USP—while Forest offers a compelling, reward-based approach for reducing screen distractions, Habitica caters to those favoring a gamified habit-building experience. Digital Wellbeing suits Android users seeking integrated digital wellness tools, and Digital Detox: Focus and Live uniquely supports focused work with its tailored messages.

The competitive analysis of the apps, exploring their features, limitations, and usage reveals crucial gaps in the current detox apps market. These challenges include retaining users, the need for tangible rewards, insufficient motivational mechanisms, limited progress tracking, a lack of community features, and a demand for customization. While these apps offer commendable features, they're predominantly virtual. Merging digital detox with tangible, real-world activities could amplify their impact. Our findings from the develop phase underscore the need to address these gaps in our forthcoming solution.

4.4 Findings from the Deliver Phase

From our interviews, we discerned pivotal habits and practices of adolescents. Our survey further probed into their social media use and its associated negative consequences. A competitive analysis laid out the pros and cons of existing apps, paving the way for the creation of our new app, JOMO, aimed at guiding adolescents towards digital detoxes. The main concept behind the JOMO application is to assist adolescents in engaging in digital detoxification in a manner that is tangible, and thus makes the process both enjoyable and conducive to their well-being. We focus on three main features—visualization of progress (develop phase), tangible rewards applicable in

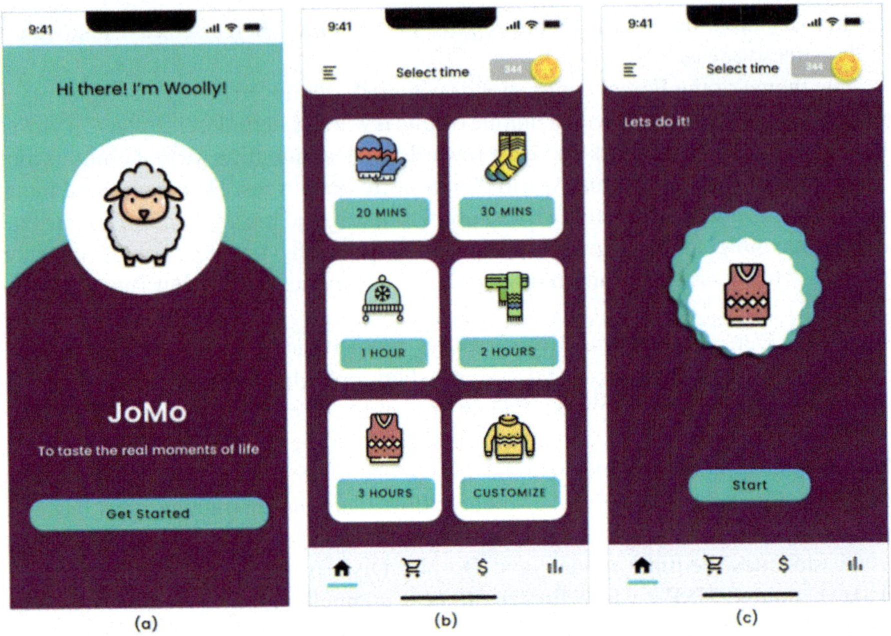

Fig. 3 **a** Splash screen, **b** home screen-different pre-set time and **c** the user selects the pre-set time he/she wants. Graphics by flaticon (www.flaticon.com)

real-life scenarios (develop phase), and motivations to reduce phubbing in a collaborative and social way (discover phase). We also introduce 'Woolly', a sheep mascot showcased in our splash screens (Fig. 3a). Wooly serves as an interactive guide to familiarize users with JOMO's features.

Visualization of progress: For mapping the progress, as a way to encourage application use, there is a need to visualize the progress with an easy-to-understand concept. After brainstorming, we finalized digital knitting as our concept to visually represent the progress and duration of time spent away from the phone. As the name suggests, our application knits a winter apparel, and each apparel is assigned a certain time for completion, with the bigger ones assigned more time, e.g., a sweater takes 3 h, but socks only require 20 min, as shown in Fig. 3b. Users may select the apparel on the basis of the time they intend to spend away from their device. Here customization plays a pivotal role, giving users control of detox time selection. Apparel size is directly proportional to both the waiting time and the magnitude of rewards, e.g., 300 virtual coins for a sweater (Fig. 3b, c). As the chosen detox period progresses, the screen displays the ongoing knitting process, complete with a countdown timer, ensuring users stay motivated (Fig. 4a).

Tangible rewards: Upon completion, rewards in the form of coins are earned (Fig. 4b), which can be redeemed in the JOMO store for real-world experiences such as trekking, hobby classes, or live events (Fig. 4c). This tangible approach aligns with

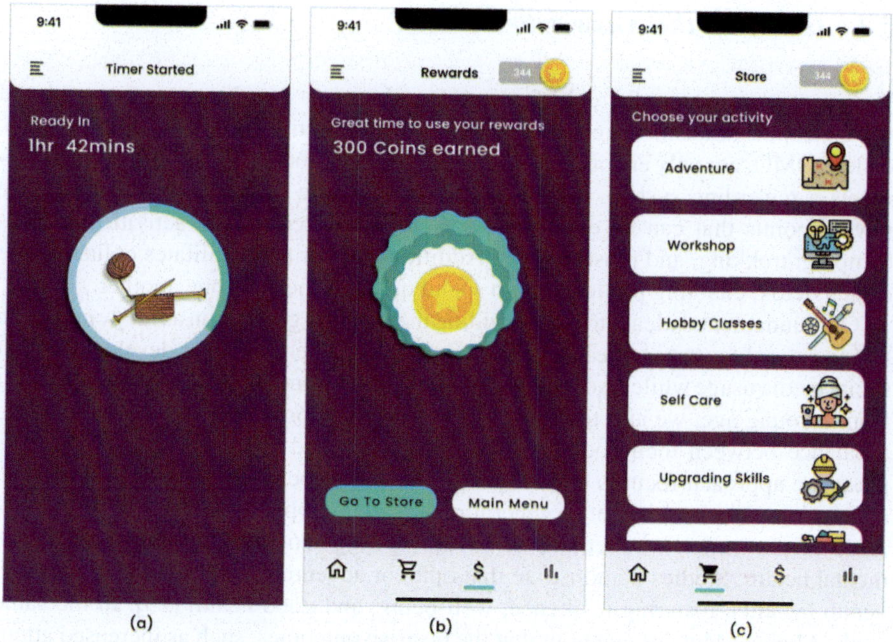

Fig. 4 **a** Animation shows the time completed, **b** The user gets points for completing the set time and **c** The user can choose the events he/she wants. Graphics by flaticon (www.flaticon.com)

our findings (as discussed in Sect. 4.2) that indicate adolescents' affinity for social gatherings and activities and overcomes the limitations with existing apps which are restricted to intangible rewards (as discussed in Sect. 4.3).

No-phubbing feature: To amplify the detox experience, particularly during social interactions, we introduced the "no phubbing" feature in our application. As reported in Sect. 4.1, interview participants reflected how their attention remained divided, even during social interactions, much to the dismay of their family and friends. For the no-phubbing feature, users pair their devices via Bluetooth, choose their detox duration together, and earn bonus points for collaboratively resisting device usage. This promotes quality, device-free interactions, and embodies the essence of our app—embracing the real world over the virtual world.

5 Discussion

In this section, we discuss our findings in relation to previous work and present the implications of our work in designing a digital detoxing app for adolescents.

5.1 Is JOMO the Answer to FOMO?

To support the pursuit of a real-life detox for adolescents, in this paper, we explored the concept of transitioning from digital to offline activities. Our innovative app, called JOMO, not only encourages participation in outdoor endeavors but also implements a rewarding system. By abstaining from phone usage, users earn valuable reward points that can be exchanged for enticing discounts on activities such as camping, trekking, and music events. Additionally, our app facilitates collaborative digital detox, enabling participants to spend quality time with one another.

Our fundamental idea is to begin guiding the younger generation as they embark on their personal journey of healthy online interactions, helping them develop efficient social media usage while also maintaining a healthy offline life. By instilling this habit from a young age, we aim to create a generation that possesses the ability to strike a balance between their social media engagement and real-life experiences. This proactive approach focuses on preventing excessive social media usage rather than attempting to address an entire generation already grappling with its consequences. After all, prevention is better than cure. Participating in outdoor activities also boosts mental health. Studies demonstrate that outdoor adventures have significant mental health benefits, fostering resilience, well-being, and good health [23]. In the same study, Michael Mutz pointed out that the positive outcomes, such as increased attentiveness, self-awareness, and reduced perceived stress, may be attributed to being in a novel environment, away from modern distractions, and in a slowed-down mode [26]. In JOMO, with the collaborative mode activated, individuals can mutually set aside their phones and engage in planned activities such as work, and assignments, or simply enjoy each other's company. This feature aims to address the issue of phubbing [27]. By utilizing Bluetooth technology, the phones can detect the presence of other users nearby, allowing for simultaneous tracking of progress. Once the designated time is completed, both users will be rewarded with points, further encouraging their commitment to digital detoxification.

Reward points earned can be later used for outdoor activities on the application as integrates a reward mechanism that fosters cooperation with various entities, including cafes, restaurants, hobby classes and workshops, trekking groups, skill development organizations, and other real-time activities that resonate with adolescents. This presents an opportunity for a business model wherein these communities stand to gain by promoting themselves on the JOMO platform, thereby expanding their audience. In parallel, adolescents benefit by receiving coupons and participating in these activities.

Unlike typical applications, JOMO does not solely serve as a time management tool aimed at deterring users from their devices. Instead, we undertake a more comprehensive approach by presenting users with wholesome alternatives for device-free moments. In doing so, we guide them on effectively integrating digital detoxification with outdoor activities, rendering the process enjoyable and uncomplicated, and ultimately enabling them to strike a harmonious equilibrium between the virtual and real worlds.

5.2 *Need for JOMO Application*

Studies show that 95% of adolescents (13–17 years old) who were questioned across all demographic groups are smartphone users, with an average consumption of 45% of the internet in daily life [5]. Studies specifically focusing on adolescents' use of social media have found that excessive use of social media impacts one's personality tremendously and for adolescents, the impact can be immense [2, 13, 39]. Excessive phone usage among adolescents also contributes to other forms of stress, particularly FOMO, where individuals feel anxious or stressed when they believe they are missing out on rewarding experiences or social interactions happening on social media platforms [1, 8]. Adolescents being the most active age group on social media, stand at a higher risk of facing the FOMO [25, 26].

While there are typical patterns of change in how adolescents respond to stress, it is evident that there exist individual variations [11]. They develop intrinsic and extrinsic coping mechanisms to deal with the stress they face in their day-to-day life. Healthy coping behaviors include problem-solving and actively seeking information, which contributes to better functioning [39]. On the other hand, poorer functioning is associated with helplessness, passivity, seeking escape, and displaying opposition [12], as also revealed in our interview study. Coping behaviors can root toward positive or negative functioning. Problem-solving and information-seeking belong to healthier functioning whereas helplessness, passivity, escape, and opposition, are associated with poorer functioning [37]. In our exploration of social media usage by adolescents, we found that excessive time spent on social media platforms, often as a form of escapism from real-life challenges, further perpetuates addiction. Additionally, the presence of online communities, while offering a sense of belonging, can also become a source of dependency and reinforcement for addictive behaviors. The prevalence of pretentious behavior characterized by inauthenticity and artificiality, amplifies the pressures and expectations associated with social media, further entrenching addiction.

5.3 *Implications to HCI Research on Adolescents and Social Media*

Our research underscores the critical necessity of developing solutions that encourage healthy social media engagement among adolescents. We advocate for the active involvement of researchers in crafting applications aimed at mitigating the adverse effects associated with excessive social media use in this demographic. This study holds substantial implications for the field of Human–Computer Interaction (HCI) by illuminating the pressing requirement to address the issue of social media addiction within the intricate fabric of adolescents' experiences.

Crucially, stakeholders including parents, educators, and mental health professionals must recognize the paramount importance of providing adolescents with guidance and support. This support system plays a pivotal role in empowering adolescents to navigate the realm of social media in a manner conducive to fostering a lifestyle that is both wholesome and balanced [32]. Furthermore, it is incumbent upon these stakeholders to formulate intervention strategies that are grounded in nature, to educate and help adolescents for a healthy social media use these strategies should encompass the regulation of time spent on social media, the cultivation of alternative coping mechanisms, and the facilitation of meaningful offline interactions. [37]

5.4 Limitations of Our Work and Opportunities for Future Work

Phase 2 of our study, the Online survey consisted of approximately 71% female participants, indicating a potential limitation in terms of gender bias. Conducting a study with a more diverse participant pool would yield more unbiased data.

Furthermore, a larger sample size user study would provide more reliable and robust findings regarding the effectiveness and use of the JOMO application on adolescents. While our study focused on the impact of social media usage, it is important for future research to consider a broader range of smartphone activities, in order to address overall phone addiction rather than solely problematic internet use. In this research, we have developed the JOMO application using insights gathered from adolescents. In the future, the next phase of this study will involve obtaining feedback and evaluations from users, which will help us make final refinements to the application's design.

Adolescents engage in various activities on smartphones, including gaming, shopping, dating, networking, watching series or shows, and communication. While our study specifically examined social media usage, future research should consider investigating the impact of these other activities to gain a comprehensive understanding of how to effectively address overall phone addiction among adolescents, rather than solely focusing on problematic social media use. By expanding the scope of research, we can develop strategies and interventions that target various aspects of phone addiction and provide comprehensive support to adolescents in managing their smartphone usage.

6 Conclusion

While social media offers numerous advantages, it also has negative effects, particularly on the younger generation who face increasing pressure to join and engage on these platforms. Symptoms of excessive phone use, such as lack of concentration, irritability, constant phone checking, and phubbing, hinder productivity. Time management apps are invaluable tools in achieving goals and maintaining a balanced life in today's digital age, but they are limited to motivating adolescents to stay away from devices. In our study, we recognize the impact of social media and excessive phone usage on adolescents through a double-diamond process and design a digital detox application called JOMO. Our work shows that it is crucial to incorporate digital detox strategies that allow adolescents to take breaks from excessive screen time, and still be socially connected with their peers and others. Through our study we provide an approach to connect digital detox with a tangible aspect, thus making it more efficient. By promoting a healthy balance between online and offline experiences, we can support adolescents in their personal growth, productivity, and overall well-being.

Appendix

A.1 Survey Questions

1. Which among the following apps are you most active on?
   (Instagram/ WhatsApp/ Snapchat/YouTube)
2. On average how many hours in a day do you spend on social media?
   (less than 3 h/3 to less than 6 h/6 to less than 9 h/more than 9 h)
3. On average how many times in an hour do you check social media?
   (Open-ended)
4. What time of the day are you most active on social media?
   (Morning 6 am–12 pm/Afternoon 12 pm–6 pm /Evening/Night 9 pm–6 am)
5. On a scale of 1–5 (5 being the highest), How important is social media in your daily life?
6. From what age did you start using social media?
   (6–9, 9–12, 12–15, 15–18)
7. Are you using social media with your real Identity?
   (Yes/No)
8. Do you have more than one account on the same online platform? If yes why?
   (Open-ended)
9. Have you felt the need to follow a certain trend on social media to stay along with friends?
   (Yes/No)

10. On a scale of 1–5 (5 being the highest), How do 'Likes' and 'Comments' on your posts affect your mood?
11. On a scale of 1–5 (5 being the highest), How seriously do you take social media platforms' suggestions on what to watch?
12. On a scale of 1–5 (5 being the highest), How seriously do you take suggestions from social media platforms on fashion?
13. On a scale of 1–5 (5 being the highest), How seriously do you take social media suggestions on which places to visit?
14. Have you brought anything from social media platforms due to the influencer's recommendation?
    (Yes/No)
15. On a scale of 1–5 (5 being the highest), how seriously do you take an influencer's advice?
16. How many influencers do you follow religiously?
    (0, 1, 2, 3, 4, more)
17. On a scale of 1–5 (5 being the highest), How bad does social media influence your sleep cycle?
18. What do you do for stress relief? (Open-ended)
19. On a scale of 1–5 (5 being the highest), How much stress is caused by seeing what your friends are doing?
20. On a scale of 1–5 (5 being the highest), How bad does social media influence your mental health?
21. On a scale of 1–5 (5 being the highest), how much does social media affect your dating life? (Open-ended)
22. What do you feel, Is social media a boon or a curse? Please elaborate on the same. (Open-ended)
23. Have you used any time management applications on your phones? If, yes name them and let us know your reason for choosing them.
24. What do you feel your ideal time management application should be like? (Open-ended)

References

1. Abel JP, Buff CL, Burr SA (2016) Social media and the fear of missing out: scale development and assessment. J Bus Econ Res (JBER) 14(1):33–44. https://doi.org/10.19030/jber.v14i1.9554
2. Agarwal S, Mewafarosh R (2021) Linkage of social media engagement with FOMO and subjective well-being. J Content Community Commun 13. ISSN: 2395-7514. https://www.amity.edu/gwalior/jccc/pdf/jun_06.pdf
3. .
4. Alanna M, Best P, Maddock A (2022) It's just one big vicious circle': young people's experiences of highly visual social media and their mental health. Health Educ Res 37(3):167–184. https://doi.org/10.1093/her/cyac010

5. Anderson M, Jiang J (2018) Teens, social media & technology 2018. Pew Research Center, May 2018. https://assets.pewresearch.org/wp-content/uploads/sites/14/2018/05/31102617/PI_2018.05.31_TeensTech_FINAL.pdf
6. Boeschoten L, van Driel II, Oberski DL, Pouwels LJ (2021) Instagram use and the well-being of adolescents: using deep learning to link social scientific self-reports with Instagram data download packages. In: The companion publication of the 2020 international conference on multimodal interaction (ICMI '20 Companion). Association for Computing Machinery, New York, NY, USA, p 523. https://doi.org/10.1145/3395035.3425185
7. Bitrián P, Isabel B, Sara C (021) Enhancing user engagement: the role of gamification in mobile apps. J Bus Re 132:170–185. ISSN 0148-2963. https://doi.org/10.1016/j.jbusres.2021.04.028
8. Brush K (2019) FOMO (fear of missing out), July 2019. https://www.techtarget.com/whatis/definition/FOMO-fear-of-missing-out
9. Cauberghe V, Wesenbeeck IV, De Jans S, Hudders L, Koen Ponnet K (2021) How adolescents use social media to cope with feelings of loneliness and anxiety during COVID-19 lockdown. Cyberpsychol Behav Social Network 24(4). https://doi.org/10.1089/cyber.2020.0478
10. Crook C (2015) The joy of missing out: finding balance in a wired world. New Society Publishers, Canada
11. Definition of FOMO abbreviation from the Oxford Advanced Learner's Dictionary fomo abbreviation—Definition, pictures, pronunciation and usage notes | Oxford Advanced Learner's Dictionary at OxfordLearnersDictionaries.com
12. Elmquist DL, McLaughlin CL (2018) Social media use among adolescents coping with mental health. Contemp School Psychol 22:503–511. https://doi.org/10.1007/s40688-017-0167-5
13. Gupta M, Sharma A (2021). Fear of missing out: a brief overview of the origin, theoretical underpinnings and relationship with mental health. World J Clin Cases 9(19):4881–4889. https://doi.org/10.12998/wjcc.v9.i19.4881
14. Hayran C, Anik L (2021) Well-being and fear of missing out (FOMO) on digital content in the time of COVID-19: a correlational analysis among university students. Int J Environ Res Public Health 18(4):1974. https://doi.org/10.3390/ijerph18041974
15. Holly S, Bibby PA (2017) Personality, fear of missing out, and problematic internet use and their relationship to subjective well-being. Comput Hum Behav 76:534–540. https://doi.org/10.1016/j.chb.2017.08.016
16. Jiuyu C (2023) Social media addiction and consequences in adolescents first. In: Lecture notes in education psychology and public media. https://doi.org/10.54254/2753-7048/7/20220823
17. Joshua SK, Cynthi J (2023) The impact of smartphone addiction on sleep quality and psychological well-being among young adults. Int J Ind Psychol 11(2):640–650
18. Aranda JH, Safia B (2018) Toward "JOMO": the joy of missing out and the freedom of disconnecting. https://doi.org/10.1145/3229434.3229468
19. Jung AL, Bright LF, Eastin MS (2021) Fear of missing out and consumer happiness on Instagram: a serial mediation of social media influencer-related activities. Mediat Cyberpsychol Behav Social Network 762–766. https://doi.org/10.1089/cyber.2020.0431
20. Kusumajati DA, Chairiyani RP, Martoredjo NT (2020) The influence of social media on human behavior in adolescents (Case study of Bina Nusantara University students). In: Proceedings of the 2020 2nd international conference on big data engineering and technology (BDET 2020). Association for Computing Machinery, New York, NY, USA, pp 102–105. https://doi.org/10.1145/3378904.3378917
21. Lipika M, Md Ghazi S, Usama AR, Pragyendu, Ritika U, Griffiths MD (2023) Mediating roles of fear of missing out and boredom proneness on psychological distress and social media addiction among Indian adolescents. J Technol Behav Sci. https://doi.org/10.1007/s41347-023-00323-4
22. Lenhart A, Duggan M, Perrin A et al (2015). Teens, social media and technology overview 2015: smartphones facilitate shifts in the communication landscape for teens. Pew Research Center, Science & Tech. Available from http://www.pewinternet.org/files/2015/ 04/PI_TeensandTech_Update2015_0409151.pdf

23. Macias T (2021) Millennials: fear of missing out (FOMO) and social media engagement. https://doi.org/10.31124/advance.14046824.v1
24. Malcolm J, Flack (2023) The role of self-esteem, depressive symptoms, extraversion, neuroticism and FOMO in problematic social media use: exploring user profiles. Int J Mental Health Addict. https://doi.org/10.1007/s11469-023-01094-y
25. Mirbabaie M, Stieglitz S, Marx J (2022) Digital detox. Bus Inf Syst Eng 64:239–246. https://doi.org/10.1007/s12599-022-00747-x
26. Mutz M, Müller J (2016) Mental health benefits of outdoor adventures: Results from two pilot studies. J Adolesc 49:105–114. https://doi.org/10.1016/j.adolescence.2016.03.009
27. Parmaksız İ (2019) Relationship of phubbing, a behavioral problem, with assertiveness and passiveness: a study on adolescents. Int Online J Educ Sci 11:34–45
28. Przybylski AK, Murayama K, DeHaan CR, Gladwell V (2013) Motivational, emotional, and behavioral correlates of fear of missing out. Comput Hum Behav 29(4):1841–1848. https://doi.org/10.1016/j.chb.2013.02.014
29. Radtke T, Apel T, Schenkel K, Keller J, von Lindern E (2022) Digital detox: an effective solution in the smartphone era? A systematic literature review. Mob Media Commun 10(2):190–215. https://doi.org/10.1177/20501579211028647
30. Sanches P, Janson A, Karpashevich P, Nadal C, Qu C, Roquet CD, Umair M, Windlin C, Doherty G, Höök K, Sas C (2019) HCI and Affective Health: Taking stock of a decade of studies and charting future research directions. In: Proceedings of the 2019 CHI conference on human factors in computing systems (CHI '19). Association for Computing Machinery,New York, NY, USA, Paper 245, 1–17. https://doi.org/10.1145/3290605.3300475
31. Sherman LE, Greenfield PM, Hernandez LM, Dapretto M (2018) Peer influence via Instagram: effects on brain and behavior in adolescence and young adulthood. Child Dev 89(1):37–47. https://doi.org/10.1111/cdev.12838
32. Sitepu Y, Harahap H (2019) Social media usage and digital detoxification on teenagers in Medan
33. Sumitha N (2023) Media addiction in adolescents. Karnataka Pediat J (PDF)
34. Walia N. Try the JOMO (Joy of Missing Out)
35. Xiuqi H (2023) The influence of social media addiction on adolescents mental states. Lecture Notes in Education Psychology and Public Media. https://doi.org/10.54254/2753-7048/7/2022943
36. Yovita SS, Harahap H. Faculty of Social and Political Sciences, Universitas Sumatera Utara, Jl. Dr. A. Sofian No. 1A, Medan, Indonesia
37. Zimmer-Gembeck M, Skinner E (2010) Adolescents coping with stress: Development and diversity. School Nurse News. https://www.researchgate.net/publication/42607713_Adolescents_coping_with_stress_Development_and_diversity/link/02e7e5251081c413dc000000/download
38. Zinaida A, Markus S, Eberle T (2018) Teenagers' reaction on the long-lasting separation from smartphones, anxiety and fear of missing out. In: Proceedings of the 9th international conference on social media and society (SMSociety '18). Association for Computing Machinery, New York, NY, USA, pp 212–216. https://doi.org/10.1145/3217804.321791
39. Ziyi Y (2023) Why adolescents are addicted to social media. J Educ Human Social Sci. https://doi.org/10.54097/ehss.v8i.4498

Work in Progress Papers

Revival of the Traditional Gond Art of Madhya Pradesh Through Modern Technologies of the Digital Platforms and UX-UI

Harsh Raj Gond and Ankita Roy

Abstract India a land of diverse cultures and traditional artforms. One of the most famous folk art worldwide known as 'Gond art' is commonly found in Madhya Pradesh, Chhattisgarh, and Odisha. Gond paintings are made by the Gond tribal community with a variety of forms and colors. The work of Gond artists is rooted with folk tales, culture and thus creates a strong story-telling element in every painting. Paintings are not just a piece of art, they are masterpieces created by renowned Gond artists practicing Gond art from a very long time. These paintings are made with intricate details by using dots, lines and shapes with vibrant colors and tell stories of animals, nature, gods and everyday lives. Gond art is considered one of the purest folk art and Patangarh of Dindori district is a village of artists, which owes its artistic legacy to the great Gond artist Jangarh Singh Shyam and other famous Gond artists. Due to pandemic and various other economic conditions, it has become a challenge for Gond artists to survive by making paintings and other crafts. This research attempts to analyze the problems faced by Gond artists in their day-to-day lives while making paintings and helps derive the possible design interventions with solutions pertaining to digital platforms and creating a seamless User Experience and Interface to aid the Gond tribal community. The data collected is through empirical and ethnographic field research, which is then analyzed to create possible design interventions leading to help the Gond tribal community.

Harsh Raj Gond and Ankita Roy contributed equally to this research.

H. R. Gond (✉) · A. Roy
Indian Institute of Technology, Hyderabad, Kandi, Sangareddy, Hyderabad, Telangana 502285, India
e-mail: harshrajgond8@gmail.com

A. Roy
e-mail: aroy@des.iith.ac.in

A. Joshi and G. R. Sim (eds.), *Proceedings of the 14th Indian Conference on Human-Computer Interaction*, Studies in Computational Intelligence 1162,
https://doi.org/10.1007/978-981-97-4335-3_6

Keywords Tribal · Gond art · Gond artists · Revival · Paintings · Digitized · Digital platform · User experience · Interface design · Culture · Empowering · Creating awareness

CCS Concepts: • **Human-centered computing → Accessibility**

1 Introduction

The Gond art is a form of painting from folk and tribal art, and it is the art that is practiced by one of the largest tribes of India and it shares its name with that tribe. The word Gond comes from the Dravidian expression, that is, Kond which means "The Green Mountains". Madhya Pradesh is the place where the Gond art is considered to be predominant, but it is also quite common in various other places such as Andhra Pradesh, Maharashtra, Chhattisgarh and Odisha. Gond art continued to be created, contained within the community for thousands of years, and nearly meeting an early demise due to invading cultures [6]. While making paintings, Lines, Dots and Dashes are used with vivid bright colors such as red, green, yellow, and blue. The paints used are natural such as charcoal, colored soil, plants sap and leaves, cow dung. Gond paintings have also been used by the Gond people as a way to record their history [5]. The Gond people believe that seeing a good image attracts good luck, a reason why they used to decorate the walls and floors of their homes with traditional motifs [14]. The Renowned Gond artist Late Jangarh Singh Shyam became the pioneer of modern-day Gond art known as Jangarh Kalam or school of art. The artist's son Mayank Shyam and daughter Japani Shyam are equally popular artists [1]. Many people have worked for the revival of art in India to create something new out of the old. Today there is a necessity to revive and regenerate the Art of Gond tribe to make their lives more comfortable and to improve their standards of living. It is our responsibility to throw some light on this Beautiful Gond tribe to encourage their work and sell more at market and inspire generations. The art is unique with its style and color, creates a storyline, very stylized figures mesmerizing the whole composition are of great value in education and cultural development of the nation.

2 Objectives

Multiple works of promotion of the Gond art form are done locally through stores, exhibitions, cultural workshops etc., but hardly anything is seen in the digital areas or any other design interventions to help the community overall. Through our empirical study it is evident that they are facing challenging times to sell their paintings especially after the Covid 19 pandemic and the sheer lack of buyer interest and commissions has created a deep void in sustaining the community practices. This research attempts to cater to the very need of this particular community and assist them with easy reach to their clients and customers. The main objective of the study is to find out the gap between Gond artists and their varied ways of the art and create a possible experiential digital market that can help them to connect to the worldwide marketplace and thus will attempt to enhance their livelihood.

3 Literature Review

Gond Art Documentary short by Light trails production provides a better overview of the Gond Artists and concern towards Identification of style by Each Gond Artist in their paintings. It also highlights the Late Jangarh Singh Shyam Tragic incident that happened in Japan to help us understand the needs of Gond artists and carefully look for things which can help them revive the art form to the fullest [7].

'Gond Culture faces threat' article by The Hindu states that the younger generation of Pardhan Adivasis, who function as bards of the Gonds preserving their myths and stories through singing, are moving away from tradition. Not just in art but their music is also suffering from and the younger generation has moved towards education and the livelihood opportunities. It has thrown up. The major problem is with the youth moving away from traditional aspects and trying to be more westernized to keep up with the new digital era [10].

'Gond artists in Madhya Pradesh work as daily wagers, vegetable vendors and laborers'. During covid and post covid too artists are struggling to feed their families. Various artists are selling vegetables on carts to earn something and other women are doing side by side work such as washing utensils and helping in household works [4]. Many artists are facing problems and their business is still suffering today [9]. The paintings are not selling and they are forced to pick other jobs to maintain their livelihood and earn bread for their families. It has been a hard time for them to create paintings, they feel tired to do so, and not able to get motivation from their heart. The pandemic has taken away patrons of Gond art leaving these tribal artists with no alternative but to work as daily wagers, vegetable vendors or manual laborers. Many artists are also trying to create environmental awareness through art to reach out to potential buyers and media [15]. Multiple atrocities related to caste discriminations of the community is also talked about in the book 'Bhimayana: Experiences of Untouchability' [13].

4 Methodology

This research on the Gond community was carried out in multiple ways and a mixed method of research was applied. In the beginning data was collected by secondary sources and overall a tribal community was understood through various available resources. Later an ethnographic field research was conducted for more than a week, where both empirical study and qualitative questions were asked to gather more information about the real issues. Interviews were conducted with the local people and the artist communities to understand the real scenarios.

4.1 Secondary Research

A secondary research was conducted on Gond art and their influence on people, culture, media and other related articles. Along with it a detailed analysis was done on their earnings, job opportunities, current ways to sell paintings from digital platforms to offline methods. The secondary research was more focused on the art style, patterns, narratives, etc. This Art form has its roots and has evolved from the floor art known as the 'digna' paintings [11]. After a thorough desk research we came to a conclusion of the topic's limitations and understood that our research would be incomplete without a direct field visit. This began with direct interactions with Gond artists of Dindori district, Madhya Pradesh and it helped us to understand the grave situation of the artists and the artforms.

4.2 Primary Research

An ethnographic field research was conducted for more than a week in Dindori District, Madhya Pradesh to gather the ground realities and information that was not easily available in the books or internet. This empirical research involved a ten day diary study and camera study on their way of life, daily practices, culture and habitual practices. More focus was given in the current scenario of making and selling paintings. To understand the process further, a questionnaire was created for interviews with the artist and their families. A total of six Gond artists were interviewed including one senior artist who coined the name 'Gond Art'. The numerous interview rounds gave us better Insights and points to focus on the right problems which can be addressed carefully. The questions were based on demographics, daily routine and lifestyle, work hours, the state of art form, connecting with clients, community practices and other issues, etc.

5 Problem Identifications and Analysis

Upon engaging in discussions with Gond Artists, it became evident that their primary challenges revolve around empowerment, financial support, identification, and education. These artists continue to grapple with persistent issues that require prompt attention. Hence, the following is a compilation of the problems confronted by Gond Artists (identities have been kept anonymous) as identified during interview sessions.

Not familiar with the digital gadgets, price of paintings are not fixed, artists are not educated, less work in the market post covid, exhibition or mela charges are costly, plagiarism of work, less support from government bodies, mediators of the gallery takes up huge amount, no personal identification of the artists, inadequate savings, awareness about Gond art, no particular system defined.

Key problems mentioned by Gond artists: no Map of Gond art and Gond Artists, no communication materials like brochure, artist catalog, website to promote Gond art, no magazines for Gond art and artists, no advertising and promotional materials, they don't have training to use phone and digital gadgets, no information regarding exhibitions, workshops happening around the globe, poor equal work opportunity and scarce regular work, meager income, inadequate support and recognition of artists and their art style, very less blogs or forums to get coverage of the community.

A competitive analysis to get an overview of how the current sale formats function in different platforms and insights were gathered. Firstly, any business model runs on commissions, so at least the middle man charges about 20–30.

6 Insights

The paintings are not sold at a fixed price. They are bound to sell at whatever amount is offered to them by the mediators of the big gallery in India and globally. Sometimes they get what they ask for once connected with clients personally. They face a lot of problems with reading and writing due to illiteracy, and language becomes a barrier while communicating with clients. Gond art is unique and each artist adds value of its own mark for their identification. They have to deal with farming and selling other things to make their living. They need funds, connections and better work scale opportunities. They have recently received a GI tag due to an acute need to preserve the art form [2].

7 Design Thinking Process

The process followed for the design research project is the Double Diamond design thinking process, which is Discover, Define, Design, and Deliver. All the data collected through field research and secondary research through online sources, forums and primary interviews, diary research conducted with various Gond artists from Madhya Pradesh was analyzed. The next step involves defining the inputs and insights collected from discover phase in the define process. The information segregation was done through affinity mapping, creating the personas and empathy mapping to better understand the pain and gain points to create the platform. The next step involved coming to a conclusion of the end result to be an application or a website, and a responsive website model was chosen due to its easy propagation. The third step includes the Design part which began with creating the wireframes, information

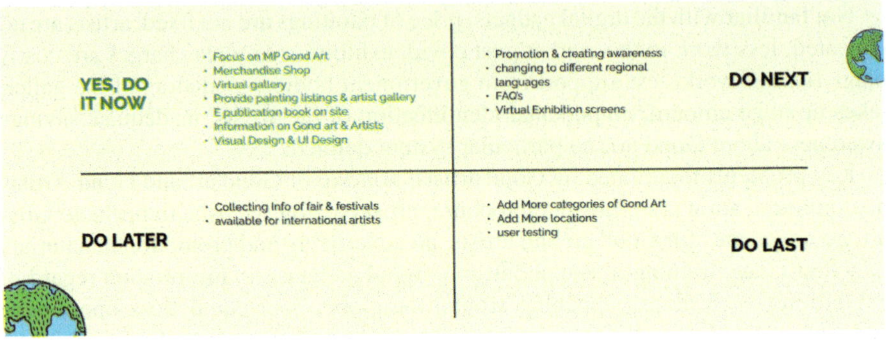

Fig. 1 Impact effort matrix chart

architecture and culture based icons, and high fidelity website prototype. Explorations were made with the design style guides, illustrations, motifs, icons and the content for the system was thought of. The later stage involved testing of the system and fixing the usability issues in the UX and UI of the high fidelity website prototype and necessary feedback was gathered, and iterations were made accordingly.

7.1 Double Diamond Process

The Double Diamond process followed during the design process while creating and ideating the digital platform which helped thoroughly during the planning, ideating and creating the mockups for the project. The Double Diamond model helped the overall project to balance divergent and convergent thinking throughout the design process. It helped avoid jumping to conclusions too quickly and ensures a more thorough exploration of the problem space, leading to innovative and user-centered solutions [8].

8 Prioritization Idea

Since the research topic involves multiple dimensions, a lot of features were thought to be included in the website, but it has its own limitations. To find effective and efficient products, one prioritization method was added and the data was analyzed and later classified into the impact-effort matrix. The goal is to stay focused and maximize time. Four categories were created, which are yes do it now, do next, do last and do later. This graph helps to bring benefit in terms of time and resource management required with the addition of a project timeline (Fig. 1).

9 The Outcome

The outcome is a Digital Web platform with the name of 'Gondkala' where Artists are being encouraged to sell paintings, build community and connect with clients directly. This will aim to provide proper information on various paintings which will be available to the customers. The platform will be a one stop platform for Gond artists to sell paintings with recognition where Artists get the full payment as there is no 3rd party involved. Platform provides 100% Authenticity of work and helps in Building connections between creators (artists) and customers, this way artists are also being identified with their own style of art. There will be an experience category option for customers to experience the virtual gallery, E-Exhibitions, EBooks, articles and blogs. There will also be a personalized merchandise store to buy t-shirts, stationary and many more items. Overall it will help enhance the user experience to both the target audience who comes to buy or for any other educational purpose. This way the platform will help revive the Gond art of Madhya Pradesh and help the artists listing their paintings on the platform and showcase their talent.

10 Designs for the Platform

10.1 Motifs for the Website

Motifs and illustrations shown in Figs. 3 and 4 that align with the various categories and enhance the overall visual appeal of the website. To lend authenticity and uniqueness, elements were incorporated taking inspiration from the Gond art painting style into the designs. All the motifs are designed digitally in Procreate software keeping in mind the style of Gond art with proper shapes and patterns. The main inspiration for illustration and UI study is from Design Code [3].

10.2 User Interface

The user interface was meticulously crafted with a focus on User Experience (UX) principles, catering to both customers and artists. Figure 2 presents the Homepage screen, featuring essential categories such as information about Gond art, available paintings for sale, exhibition details, and customer reviews, all contributing to an authentic and immersive browsing experience. Exhibition section offers digital interaction with exhibitions, e-books, and paintings.

The "About Gond Art" page comprehensively covers the rich history of Gond art and incorporates an interactive map of Madhya Pradesh, enabling seamless navigation. Additionally, an online store is integrated, allowing users to purchase artworks, while the E-exhibition section offers digital interaction with exhibitions, e-books, and paintings.

Fig. 2 UI screens: homepage, about, store experience

The Artists page, a dedicated section that showcases a curated listing of artists. This page provides users with a glimpse into the world of each artist, presenting details about their background and artistic journey. Additionally, visitors can explore and purchase the artists' paintings that are available for sale, immersing themselves in the diverse range of artwork on offer.

In general, this platform provides an extensive range of painting listings and educational information to assist users in understanding Gond art. Furthermore, the platform is designed to be highly intuitive, allowing users to easily navigate between different categories.

11 Conclusion

Post covid and the impact of social and political changes in the country have led to reduction in the work of Gond Artist. There is an urgent need to revive and regenerate the Gond art in the country to encourage and empower the local Gond artists and overall community to create more of such art forms and inspire generations to come. The Gond art of Madhya Pradesh is quite unique and remarkable which needs a help to uplift the art and the artists living there. Some earlier work has been done in the similar field but no such permanent help has been given to fix their problems of money and livelihood. Looking at their problems it seems there should be enough wide coverage and popularity of Gond art and the artists. In order to address these challenges, we are working closely with Gond artists of Madhya Pradesh to create and generate this particular platform to help in the sale of their paintings but also serve as a means to promote and raise awareness about the unique art form and support the tribal people .

Some examples where the references have been taken from, [6] an article on Gond art unpacking greatness, [5] an article about the art of Gond, Handmade in India, article about Madhya Pradesh's Gond paintings and why each piece is a masterpiece [14], more resources collected from the World of Gond art by Bose [1], Gond art short documentary film portraying the life of Gond artists [7], Gond culture faces threat [10], More insights found from Quarantined Art, Survival at stake [9], Goan connection shows Gond artists in Madhya Pradesh work as daily wagers, vegetable vendors and MGNREGA labourers [4], Trends in the Gond Aesthetics and Narrative Styles in Graphic Novels and beautiful comics [11], Creating environmental awareness through Gond art showed a great amount of insights for the finbal study [15], article on need to preserve this ancient tradition [2], some design thinking articles and blogs [8], article on Experiences of Untouchability [13] some examples collected from UI Design handbook illustrations [3] and an article on displacement of tribals impacting art [12].

The envisioned platform will not only contribute to the artists' livelihoods but also provide them with the encouragement and recognition they deserve. By establishing this platform, we aspire to enhance their visibility and enable them to establish their individual identities within the art community. This research paper explores the challenges encountered by Gond artists and addresses the authentic procurement of Gond artworks. The authors present a comprehensive solution centered around an intuitive portal designed for easy buying and selling of paintings by both artists and customers. The user interface is intentionally crafted to be simple and visually appealing, aligning with the essence of Gond art. Additionally, an efficient messaging platform is incorporated to facilitate swift navigation. In essence, the authors prioritize the perspectives of Gond artists in the design of the portal's user interface, aiming to enhance accessibility and efficiency to cater to their specific needs. They contend that this proposed solution holds the potential to significantly and positively transform the global landscape of selling and buying Gond art while simultaneously serving as an educational resource to promote awareness of this art form worldwide.

References

1. Bose SK (2022) World of Gond art. Number 4. https://www.deccanherald.com/special-features/world-of-gond-art-1107236.html
2. Chaturvedi V (2023) Gond art is laudable, but acute need to preserve this ancient tradition. https://timesofindia.indiatimes.com/city/bhopal/gi-tag-for-madhya-pradeshs-gond-art-is-laudable-but-acute-need-to-preserve-this-ancient-tradition-experts/articleshow/99570278.cms?from=mdr
3. Design Code (2021) UI design handbook illustrations. https://www.figma.com/community/file/1092769222074462910/Design-thinking,-Double-diamond-&-UX-tools/Design-thinking%2C-Double-diamond-%26-UX-tools
4. Goan Connection (2021) Gond artists in Madhya Pradesh work as daily wagers, vegetable vendors and MGNREGA labourers. https://en.gaonconnection.com/gond-art-paintings-madhya-pradesh-mgnrega-labourers-tribal-artists-gondwanaland-poverty-development-india/
5. Housenama (2017) The art of Gond, handmade in India. https://housenama.com/blogs/handmade-in-india/gond-art
6. Jaya He GVK New Museum Store (2019) Gond art unpaacking greatness. Commun ACM 50. https://jayahegvknewmuseumstore.wordpress.com/2019/09/01/gond-unpacking-greatness/
7. Light Trails Production (2022) Gond art short documentary film. Number 5. https://www.youtube.com/watch?v=uY6igH_Njhw&t=6s
8. Piquero M (2022) Design thinking, double diamond & UX tools via Figma. https://www.figma.com/community/file/1092769222074462910/Design-thinking,-Double-diamond-&-UX-tools/Design-thinking%2C-Double-diamond-%26-UX-tools
9. Saha N (2020) Quarantined art, survival at stake. Technical report 7. http://tehelka.com/quarantined-art-survival-at-stake/
10. Singh SH (2022) Gond culture faces threat. Number 6. https://www.thehindu.com/news/national/telangana/Gond-culture-faces-threat/article60254207.ece
11. Singh D, Roy A (2023) Trends in the Gond aesthetics and narrative styles in graphic novels. https://doi.org/10.1007/978-981-99-0428-0_10
12. The Citizen (2023) Displacement of tribals is also impacting art. https://www.thecitizen.in/life/displacement-of-tribals-is-also-impacting-art-969470
13. The Comics Journal (2012) Experiences of untouchability. https://www.tcj.com/reviews/bhimayana-experiences-of-untouchability/
14. Times Travel Editor (2022) About Madhya Pradesh's Gond paintings and why each piece is a masterpiece. Number 3. https://timesofindia.indiatimes.com/travel/destinations/

about-madhya-pradeshs-gond-paintings-and-why-each-piece-is-a-masterpiece/articleshow/75626096.cms
15. Vijayakumar V (2020) Creating environmental awareness through Gond art. https://www.thehindu.com/news/national/tamil-nadu/creating-environmental-awareness-through-gond-art/article32058656.ece

Touchtalk: Enhancing Messaging in Smartphones for Visually Impaired

Bindi Shah and Mrunmai Abhyankar

Abstract Messaging applications have become integral to modern communication, enabling people to connect and share information. Despite efforts to design gadgets for visually impaired (VI) individuals, existing interfaces remain inaccessible. This research paper is focused on the hurdles that VI users experience while transitioning to devices with smooth screens for basic messaging tasks. An extensive review was conducted by authors for 17 papers covering various keyboard layouts and usability features in smartphones for messaging applications. The authors also conducted a preliminary study with 18 participants aged 12–48 to understand their pain points and needs. The paper reports implementing a 9-key-based keyboard layout in a messaging application. The layout is similar to the multi-tap system used on numeric keypads of feature phones, which had several letters associated with each number key. Additionally, the proposal includes a phonic instruction to make it easier for visually impaired individuals to learn and use this keyboard layout. The authors have created the application and are currently in the initial stage of conducting user testing. They aim to assess the accuracy of the layout in the near future. Thus, the authors aim to reduce cognitive effort, simplify the learning process, and improve the overall texting experience for users with visual impairment via their application - Touchtalk.

Keywords Accessibility · Messaging · Touch screen · Usability · Visually impaired

CCS Concepts:• **Human-centered computing → Accessibility**

B. Shah · M. Abhyankar (✉)
University of Texas at Austin, Austin, TX, USA
e-mail: abhyankarmrunmai@utexas.edu

B. Shah
e-mail: bindi276@utexas.edu

A. Joshi and G. R. Sim (eds.), *Proceedings of the 14th Indian Conference on Human-Computer Interaction*, Studies in Computational Intelligence 1162,
https://doi.org/10.1007/978-981-97-4335-3_7

1 Introduction

At least 2.2 billion individuals worldwide have a near or far vision impairment [6]. This accounts for about 28% of the world's population [20]. Smartphones have evolved significantly as technology has advanced. When the world shifted from keypad-based phones to touchscreen devices, the visually impaired struggled without the tactile feedback of physical keyboard buttons. Finding icons, entering buttons, and comprehending the displayed messages are only a few of the significant obstacles.[22] India is home to about 20% of the world's blind population [3], yet braille literacy in India is merely 1% [5]. Hence, even if a solution is developed using a braille keyboard, the majority of the visually impaired may not be able to benefit from it.

Despite various new gadgets being designed with visually impaired users in mind, the interfaces fail to provide accessibility since the target users are unable to grasp the new technology and interface. Since smartphones have smooth screens displaying all the content, it is nearly impossible for the visually impaired to read content from the screens without assistance [23]. Many smartphone manufacturers have tried including other accessibility features like magnification, colour contrast, and text-to-speech embedded programs [7]. However, these features should be easily discoverable. Although some of these features have helped a lot of the target group, many have yet to benefit from them [27]. In an era where smartphones have become ubiquitous tools for communication, information access, and daily life, closing this accessibility gap is vital to ensure equal participation and opportunities for all.

The accessibility and usability of smartphones for the visually impaired have been a key topic in research, and much work is being done in this area. On interacting with the blind people, the authors understood that they had to memorize the layout of the phone's keypad, which is quite similar to conventional phones but with two extra keys, send and cancel. After mastering these keys, the users could use most of the phone's functions even when they could not see the display. Based on the surveys conducted by the authors, it was observed that although they were well-versed with button-based phones, they preferred smartphones over button-based devices. Most phones now have touch-based displays which necessitate visual navigation, excluding the visually challenged.

This study examines how touch-based phones might be made more accessible to the visually impaired. The usability of smartphones for VI users and different keyboard layouts were studied and a 9-key multi-tap layout was selected. A prototype was created to replace the phone's user interface and give a simple method to enter text on a touch-based smartphone. The contribution of this work is an Android-based application that illustrates the innovative user interface, which allows the visually impaired to engage with a smartphone with ease. The layout is similar to the multi-tap system that was utilized on the numeric keypads of older feature phones, where each number key was associated with multiple letters. Moreover, the proposal includes a phonetic guide to improve accessibility for visually impaired individuals, assisting them in both learning and using this keyboard layout. The operation is based on Talkback, which allows the user to navigate through voice assistance.

2 Related Work

The field of research on mobile applications and their accessibility for the visually impaired is relatively new and emerging. The study delves into multiple facets of mobile texting for individuals with visual impairments, including the development of novel keyboard layouts. These identified areas are systematically categorized into the following groups:

2.1 Braille-Based Keyboard Layouts

Braille is a tactile system made up of raised dots that help the visually impaired communicate and read. It provides access to written data through the sense of touch. In today's digital age, there are numerous smartphone applications designed to empower individuals with low vision by enabling them to type in Braille.

"BrailleTouch" [24] was an application developed to allow visually impaired users to access the keyboard using Braille script. It claimed that since many VI users are familiar with Braille, it would be easier for them to type using the Braille layout. However, the results were not as promising as expected. The users could not find the exact location of keys on the smartphone due to the smooth screens. The application was further developed in [11]. The keyboard featured a six-key layout, with each key representing two distinct Braille characters. To input Braille characters, users simply tapped the touchscreen, and the system provided immediate audio feedback, enhancing the user experience and ensuring accurate text entry. However, users who are not familiar with Braille might find it challenging to adapt to this input method, potentially limiting its accessibility to a broader audience.

Researchers in [14] created a customised Braille-based keyboard that uses VoiceOver to provide feedback on what the user enters. However, the application was only limited to SMS service and was tested on only 7 users leading to insufficient data to draw conclusions. In [8], the authors developed a single-tap Braille text entry system for smartphone screens which were divided into two columns and three rows. Different swipe directions conveyed different meanings. The system was simply reliant on swipe and tap actions, eliminating the need to locate the position on the screen. Despite the innovative approach described there are drawbacks to consider. Firstly, the limited sample size of only three participants in the prototype testing raises questions about the system's overall effectiveness to a wider user base. Additionally, the system's performance in real-world scenarios, such as noisy or distracting environments, has not been explored, leaving potential usability issues unexamined. In the next study presented in [24], the authors engineered a tactile Braille keyboard specifically tailored for smartphones. This keyboard aimed to facilitate proficient text input on mobile devices, catering to VI users with Braille proficiency. Notably, Google has implemented a dedicated Braille keyboard [4] into its smartphone operating system, boosting accessibility and inclusion for visually impaired users. As depicted in the Fig. 1, is a braille-based keyboard by Google.

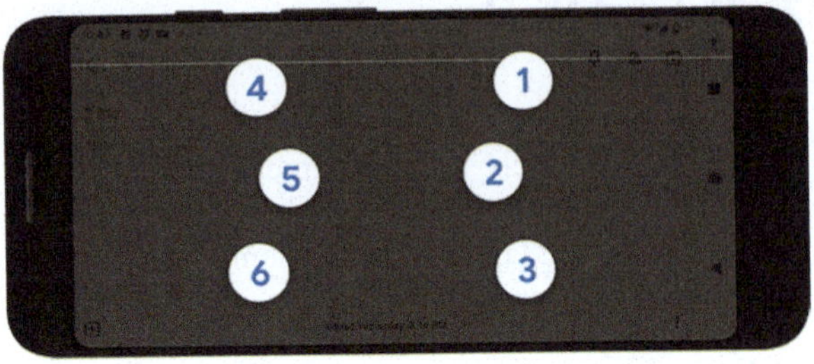

Fig. 1 Braille keyboard in smartphone [4]

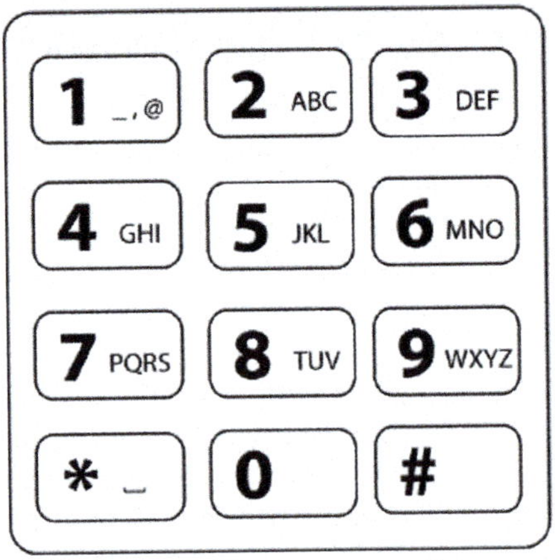

Fig. 2 12-key-based keyboard layout [28]

2.2 12-Key Based Keyboard Layout

Smartphones use a variety of keyboard layouts, although the QWERTY and 12-key layouts are the most common. The QWERTY keyboard layout, which is similar to the typical computer keyboard, has a full set of alphanumeric keys organised in a grid. The 12-key layout, on the other hand, is a simplified version that uses a numeric keypad with each key representing numerous letters and characters. These two keyboard layouts are the most popular choices in smartphone design, providing a variety of user preferences and input patterns. The 12-key-based keypad layout is represented in Fig. 2 as mentioned in [28].

In a notable variation of traditional keyboard layouts, an innovative design known as "Stick," introduced in [12], re-imagined the user interface by mapping the characters from the familiar QWERTY keyboard onto a compact 12-button-based layout, similar to the keypads found on older mobile phones. While multi-tap layouts may seem time-consuming and frustrating at first glance for some, small screens with limited space, multi-tap layouts can maximize screen space by using fewer buttons, making it easier to navigate on smaller screens.

Fig. 3 9-key based keypad layout [2]

1	2ABC	DEF3
4GHI	5JKL	MNO6
7PQRS	8TUV	WXYZ9

2.3 *9-Key Based Keyboard Layout*

In the research paper described in [26], A 9-key layout was created exclusively for smartphone screens, with a view of enhancing the typing experience for those with visual impairments (VI). The primary goal was to reduce the use of external typing devices, enabling direct text input on mobile devices for greater convenience and efficiency. What makes this strategy so interesting is the good response it collected from the research study participants. Their enthusiasm and suggestions highlighted the layout's potential to improve accessibility and usability for VI people. However, it is important to note that, surprisingly, there has been a significant absence of future study efforts in this specific field, leaving a significant opportunity for future investigations and advancements in accessible technology. Figure 3 represents a simple 9-key-based layout in phones.

2.4 *VoiceOver Based Inputs*

Many smartphones use the feature of voice assistance to help the visually impaired access smartphones easily. An application called Voisee presented in the paper [18] was developed to bridge the communication gap between visually impaired and deaf users. It used Speech to Text and Text to Speech in order to convert the input and output formats. However, it was an SMS-based application that required a SIM card to send and receive messages. Moreover, the users reported that the application lacked an interactive interface. The authors of another paper [29] developed a voice-based application, called Voicemoji. The application provided a voice-based keyboard for entering emojis for the visually impaired. It was tested on 12 participants and yielded good results. However, the application was limited to iOS users and was tested on only 12 participants.

Another simple application was proposed for enhancing calling and messaging for the visually impaired in [25]. The authors of the paper incorporated Speech to Text and Text to Speech in order to communicate via SMS/call. Visually impaired users were able to talk into the application, and their voice was transcribed and delivered to another user who read/listened to the message. In [15], the architecture and working of Android and how the Speech-to-Text feature works for transmitting messages from one device to another was studied. Different steps of the process were studied in depth and tested via a prototype built by the authors. In the proposed prototype, a QWERTY keypad was used along with controls like power off, volume up/down, call, and end call for the speech-to-text functionality. However, the proposed prototype was not tested among the user group and did not tell how accurate the system is.

2.5 Other Keyboard Layouts

Another experiment [9] was done by dividing the screen into eight wedge shapes replacing the QWERTY keyboard layout that used two-finger interaction with an audio feedback. The typing system turned out to be about twice as fast as compared to VoiceOver.

Researchers in [13] discuss a navigation method designed to assist blind users in text input on mobile devices while reducing cognitive load. Eight blind users were involved in the prototype's design over three weeks, and five users participated in a 16-week study period. It divides the alphabet into five lines, each starting with a different vowel, and uses joystick-like navigation to select letters. This method provides audio feedback on the selected letter before finalizing the input. Users can navigate cyclically both vertically and horizontally. The solution was based on the physical multi-tap phone that uses specific letters in a group to help users identify the location of letters faster.

2.6 Usability Features

In [16], the authors evaluate the usability of three Android messenger applications: WhatsApp, Skype, and GO SMS Pro, chosen based on user preferences. The study examined primary tasks like chatting, file transfer, profile management, and updating contacts. Cognitive Walkthrough was used for evaluation, revealing issues like difficulties in selecting multiple emojis, lack of file transfer confirmation, and ineffective search functionality. Since texting in mobile applications can be a tedious task for the visually impaired, different keyboard layouts available were analysed in [19] and an ideal layout was proposed but nothing was developed on the prototype further.

2.7 Other Applications

In [21], the authors presented an Android application that assists the visually impaired in performing simple assistive tasks like SMS, emailing, outdoor navigation, and schedule management. The application used speech-to-text to get voice commands and Natural language processing to assess the command and perform the appropriate tasks. The application was capable of supporting three languages, making it more inclusive for people who were not well-versed in English. In article [10], D. Burton provided a review of two cell phone-based screen readers designed to assist individuals with visual impairments in accessing and using mobile phones. The review primarily focused on the accessibility features and usability of these screen readers. The review focused on only two specific screen readers, potentially excluding other available options. This limited selection may not provide a comprehensive view of all available solutions. An educational chatbot was developed in [17] that consisted of an Artificial Intelligence-based model that uses Natural Language Processing to input queries and questions for the VI users, process them, and return answers in the form of voice or text. The system used MediaWiki API or AB Library APIs for retrieving the information. Though the efforts were made, the information can not always be reliable since AI-based systems are prone to errors.

The literature review examined various keyboard layouts and thoughtfully outlined their respective advantages and drawbacks. Several issues were identified and are summarized below for clarity.

- While commonly used, the QWERTY keyboard [12] faces a fundamental obstacle - the near proximity of keys, which needs frequent navigation through a cluster of keys to reach the correct one. This proximity frequently results in unintended key presses, resulting in typos and a less-than-ideal user experience.
- The Braille keyboard [11], on the other hand, while created with inclusivity in mind, takes up the entire screen while in use. This full-screen employment prevents users from doing anything else on their smartphone while typing. This constraint can be inconvenient for users who need multitasking skills or who want to be aware of other operations or notifications while writing text.

3 Motivation

The review of existing literature enabled to realize that limited work has been done on 9-key and 12-key keyboard layouts [12, 26]. It was found that only a few prominent papers have picked the layout as their research focus. This caught the author's interest in exploring the layouts further. The review indicated reasons for the absence of continued research in the 9-key layout for visually impaired users. It stems from several factors like limited resources, rapid pace of technological change, competing research agendas, and potential ethical issues. However, the authors believe that these difficulties can be addressed by taking a proactive approach. Securing funds, increasing awareness among stakeholders, and seeking partnerships with institutions and accessibility experts are all critical initiatives. Engaging visually impaired people directly for feedback, publishing findings, and pushing for legislation changes can also be ways to overcome these obstacles. The authors later sought an alternative solution to make their knowledge and ideas come to life. Their inspiration came from the design of button-based phones, prompting them to explore the integration of a user-friendly multi-tap layout in smartphones. This approach aimed to simplify text input for users while capitalizing on the familiarity of traditional keypad-based devices.

4 Methodology

With the motivation to develop a usable solution for the visually impaired, the authors talked to the target group to understand their problems and difficulties further. For that, the authors contacted several visually impaired users via social media as well as scheduled visits to blind schools.

4.1 Participants

The authors conducted online interviews, circulated online questionnaires as well as personally visited blind schools. They sought prior permission from the school principal, parents, and guardians of the minor students A total of 18 participants (Mean = 20.94, SD = 13.37) were involved in the interview process. 67% of the participants belonged to the age group of 13–18 years and were actively using smartphones. The key rationale for selecting this age group was the generation's rising use of smartphones. Smartphones are more attractive and familiar to the younger generation than to older generations. The remaining 33% participants were adults, whose information was obtained via online surveys circulated through social media. Figure 4 represents the classroom scenario when the authors surveyed the blind students.

It was observed that the adults owned a smartphone and were well-versed in its usage. Of all the users surveyed and interviewed, about 44% participants were totally blind, 28% participants were able to identify light and shapes and the rest 28% were low visioned, i.e. they had limited visual acuity. Around 83.3% of users were excellent with Braille and learned Braille at a very young age. The demographic data of the participants is presented in Table 1.

P01 is the youngest participant. She has been partially blind since birth and started formal education through Braille at the age of 4. P05 is a 14-year-old student. She lost her vision recently due to an accident. Despite the circumstances, she managed to learn Braille quickly within a year. P06, a retired professional, started to lose

Fig. 4 Interaction session with blind students

Table 1 Demographic details of the participants

Name	Age	Gender	Education	Visual difficulty level	Impairment since
P01	12	Female	8th grade	Blind—can see light and shapes	Birth
P02	16	Male	10th grade	Totally blind	2 years old
P03	28	Female	—	Blind—can see light and shapes	17 years old
P04	22	Male	12th grade	Blind—can see light and shapes	Birth
P05	14	Female	9th grade	Totally Blind	13 years old
P06	61	Male	BTech	Low visioned—limited visual acuity	55 years old
P07	13	Female	8th grade	Blind—can see light and shapes	Birth
P08	13	Female	6th grade	Low visioned—limited visual acuity	7 years old
P09	48	Female	12th grade	Totally Blind	Birth
P10	34	Male	D Ed.	Totally Blind	Birth
P11	21	Female	12th grade	Blind—can see light and shapes	Birth
P12	13	Female	7th grade	Totally blind	Birth
P13	15	Female	7th grade	Blind—can see light and shapes	1 year old
P14	13	Female	6th grade	Low visioned—limited visual acuity	Birth
P15	14	Male	7th grade	Totally Blind	Birth
P16	13	Female	6th grade	Low visioned—limited visual acuity	2 years old
P17	13	Female	6th grade	Totally Blind	Birth
P18	14	Male	7th grade	Blind—can see light and shapes	Birth

his vision 7 years ago due to diabetes. This lead to anxiety and stress, affecting his condition more. P09, aged 48 years, is blind since birth. She is a mathematics teacher by profession and has completed her education in the same school. She was fluent in performing simple tasks on her smartphone using Talkback. P16 is a sixth-grade student. She was diagnosed with low vision at the age of 2 when her parents noticed her facing difficulty recognizing objects.

4.2 Procedure

All the interviews took place in the years 2022–23. Various visits were scheduled to blind schools. Figure 5a is a photograph taken live during an interview held with one of the respondents. In addition to visiting blind schools, the authors connected with visually impaired users through social media and performed online interviews to acquire additional insights and learn more about the users' issues. The interviews were conducted through Google Meet and Zoom calls. Figure 5b is a screenshot of the online meeting conducted with one of the participants.

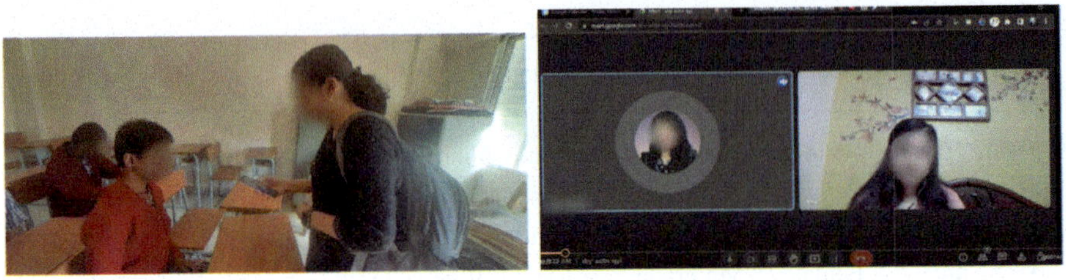

(a) The author in discussion with a participant (b) The author conducting personal interview remotely

Fig. 5 Interviews conducted by the authors

Table 2 Interview questions

Sr. No.	Question
1	What is your level of visual difficulty?
2	Since when have you been visually impaired?
3	What is your educational qualification?
4	Which messaging applications do you use on your phone?
5	Do you need assistance while messaging?
6	What frustrates you the most while using messaging applications?
7	Have you used accessibility features?
8	Have you used a button based phone?
9	What type of phone do you prefer to use?
10	While using smartphones, which text input method do you use?
11	What modes of communication do you use while replying to messages? (For example: text, images, emojis, locations, gifs, audios, video sharing, documents)
12	How well do you understand the accent and speed of the voice assistant?

The average length of the interviews was roughly 10 min, with the extra theme being general questions to further encourage the respondent to think freely about their challenges in life. The questions were arranged in such a way that the interviewees felt at ease before going on to more possibly sensitive topics. The questions were framed to be open-ended and chosen to make it easier for them to answer. On a scale from 0 to 10, where 0 indicates complete dissatisfaction and 10 signifies complete satisfaction, participants were asked to rate their level of understanding of the given question. This process assisted in conducting a quantitative assessment of opinions and perceptions of the respondents' sentiments. Further, it provided a structured framework for gathering data and analyzing patterns and responses, ultimately contributing to a comprehensive evaluation of the various accessibility features in smartphones. The questions asked in the survey are presented in Table 2.

Following the interviews, the purpose was to evaluate them to determine the users' requirements and goals. Because the interviews were semi-structured, the data analysis began with a transcription of the recordings, followed by highlighting parts containing all of the concepts, ideas, and conclusions in each interview. Before compiling the data, the findings were organised into points with a little notation on who stated what. During the interviews, several topics were mentioned by multiple people and therefore regarded to be a common concern to consider.

5 Results

After conducting online interviews and visits to blind schools to obtain a greater understanding of the visually impaired and the obstacles they encounter when using smartphones, the target group's comments were listed and carefully examined to make inferences and gather insights. From the intensive surveys conducted, the following key insights were drawn:

- When asked if they required help when using the messaging apps, 33.3% of users claimed to be self-sufficient, while 44.4% needed help quite frequently. The rest claimed to not be able to independently operate a smartphone.
- The majority of the participants were familiar with TalkBack/VoiceOver [1]. However, a few of them had already memorised the location of applications, while the others required regular assistance when using the phone.
- Approximately 84.7% of the participants interviewed have reported using VoiceOver/TalkBack to respond to text messages. There has been extensive research and development in the area of VoiceOver-based messaging applications [15, 18, 25, 29] designed to assist individuals with visual impairments.
- The participants emphasized their dependence on the VoiceOver feature. However, one of the participants (P10) also stated that is it not 100% reliable since it provides limited context, causing a communication gap between the users and the smartphone.
- A few users expressed their frustrations. One of the participants stated that the absence of physical buttons makes understanding the smooth screen difficult without a voice assistant. Another user stated, 'The Read More option - which is used to expand long texts when requested—in the messaging application Whatsapp is bothersome since it interrupts the existing message'.
- Approximately 58.3% of the participants were ignorant of or had not used the accessibility functionality built into their devices. Two other users reported that the accessibility features were ineffective to them.
- On a scale of 0–10, all participants who have used accessibility features of smartphones rated the ease of accessibility features as 7 or above.
- Around 77.8% of respondents claimed to have used button-based phones and all of them were fairly satisfied with the convenience of the keypad-based phones. As mentioned by the researchers of [12], even though a multi-tap layout can be time-consuming, it is easier to navigate for the visually impaired.
- Even though such a large portion of users stated that button-based phones were convenient to use, they still preferred smartphones.

- Moreover, 7.7% of them used a Braille keyboard for their text input needs. This aligns with the fact that Braille literacy in India is relatively low, estimated to be around 1% according to literature [3]. The remaining 7.6% rely on continual help when texting.

Frequent patterns observed during the interviews pointed to a fundamental issue that required careful consideration. The understanding of these trends was derived by analyzing findings from both online surveys and in-person visits.

- Despite being familiar with Braille, users did not use the Braille keyboard.
- The target audience claimed to be highly reliant on smartphones' voice assistant features.
- Users said that the smartphone was useless without TalkBack.
- Furthermore, participants stated that they were familiar with button-based phones and would prefer the tactile feedback of button-based phones on smooth screens of smartphones.

The survey results further contributed to the ideation of a simple keyboard that can enhance accessibility for the visually impaired.

6 Proposed System

The authors proposed a 9-key-based multi-tap keyboard layout in a messaging application, stating several compelling reasons for doing so. The optimal use of screen space, the potential reduction in the number of required keystrokes for text entry, and the introduction of auditory feedback to improve the user experience are just a few of the important benefits. The authors proceeded with the creation of a messaging application to bring this vision to reality. They selected the versatile Flutter framework, utilizing the Dart programming language as their primary tool. This software is intended to be compatible with Android phones running versions higher than 4.1 (Jelly Bean), ensuring a large user base. The keyboard takes up one-half of the screen here, while the upper half accommodates the application's user interface. This planned layout serves to improve the user's typing experience as well as their interaction with the application's content, with an aim to provide an intuitively efficient messaging platform. The application is named 'TouchTalk' by the authors. Figure 6 represents a visual representation of the proposed solution.

The key attributes of the proposed system are as follows:

- The 9-key layout consists of a grid of nine keys, each associated with a specific set of letters or characters. This layout is consistent, with the keys in the same positions every time. Compared to a full QWERTY keyboard with many keys [15], a 9-key layout offers a smaller number of key options. This can make it easier for users to locate and press the desired keys without the mental effort of deciphering a larger array of keys.

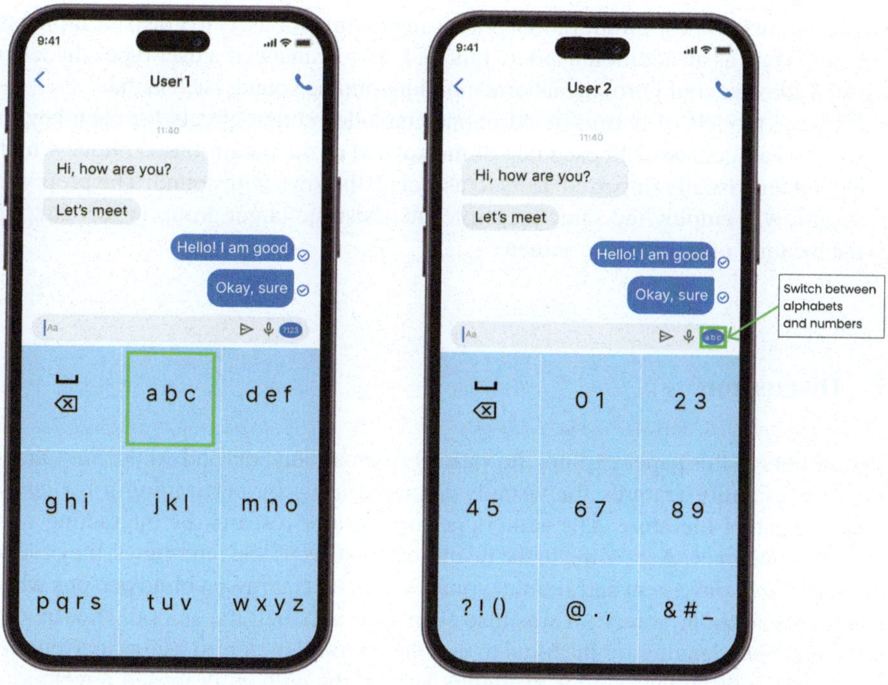

Fig. 6 Touchtalk—a messaging app with a 9 key-based keyboard layout

- The minimalist design layout allows fewer distractions on the screen. As mentioned and claimed in paper [26], users can focus on the content that they are typing without navigating a more complex keyboard interface.
- The user will eventually possess a good understanding of the letters present on each button due to practice, making it effortless for them to quickly identify and locate the desired letters. For instance, if the user is aware that key 2 contains ‘a’, ‘b’, and ‘c’, and wants to input the letter ‘c’, the user does not have to navigate everywhere and can just look for key 2.
- The screen is often partially obscured by the user’s hands or thumbs. This physical barrier may further protect sensitive information from prying eyes.
- The use of the proposed 9-key-based keyboard may feel more natural and comfortable for blind individuals switching from feature phones with physical keypads to smartphones, easing the learning curve.
- The application also has the potential to reduce the keystrokes and time required to find the desired key.
- The layout provides a larger area on the screen for a group of letters rather than allocating small spaces to each letter, which can be cumbersome for the visually impaired to use.

- The authors propose an additional phonic instruction for users to grasp the alphabet being typed as an additional aid to Talkback. For instance, if a user types the letter 'A,' Talkback could provide a corresponding phonic sound, such as 'ah.'
- Previous models of button-based phones included sound effects for each key. A sound is associated with each key of the keypad of the button-based phone which helped the visually impaired users to associate the keys with sound. The proposed system will employ the same principles to assist the target group in recognizing the location of keys on the screen.

7 Discussion

The authors of this paper explore the various keyboard layouts and other smartphone assistive usability features for visually impaired users by performing a review of related existing literature. The authors performed user research by interacting with the target audience. According to the results of the surveys and interviews, they found the need for a convenient and simple typing system that can assist blind persons when using a smartphone screen. It was also seen that, although the smooth smartphone screens are challenging for the blind to use, the VI's still preferred using smartphones owing to the different features available. Hence, the authors proposed a technique intended to reduce cognitive load in order to provide easy access to messaging. With prolonged use of the keyboard layout, the users will be able to determine the positions of buttons based on their muscle memory and will be able to type faster. The authors' recommended 9-key multi-tap layout has the following advantages: minimalist design, less cognitive load, and a quicker rate of typing.

However, it is also crucial to acknowledge the study's limitations. One limitation is that the present system does not allow multiple languages and is restricted to only one. To do this, the author intends to provide a future solution by assisting users in locating key positions by introducing a thin frame of aluminium or silicon linked to the phone casing. The guiding frame will have a 3×3 grid arrangement that corresponds with the edges of the buttons on the screen, acting as a guiding surface for the blind to locate the buttons. In this study, user testing was not conducted. Another major limitation of this work is the absence of user testing for the proposed keyboard. The application can be tested via different techniques with users of different demographic distributions which can provide promising results.

8 Conclusion

This research paper delves into the challenges faced by visually impaired individuals when engaging in mobile messaging and seeks to address these issues comprehensively. The authors propose a solution that revolves around a 9-key multi-tap keyboard layout reminiscent of traditional physical multi-tap phones. This layout aims to provide an intuitive and efficient messaging platform, offering users a swifter means of navigation between the keys. They are actively working on the prototype and are in the initial stage of conducting user testing. In brief, the authors aim to prioritize the viewpoints of visually impaired users while designing a smartphone interface that offers enhanced accessibility and efficiency to meet their distinct needs. The authors believe that this solution has the potential to bring a significant and positive transformation in the use of smartphones by visually impaired individuals.

References

1. 2014. What is Google TalkBack?—androidcentral.com. https://www.androidcentral.com/what-google-talk-back. Last Accessed on 8 Oct 2023
2. 2018. Design of Smartphone 9-Key Keyboard Based on Spelling Rule of Pinyin—link.springer.com. https://link.springer.com/chapter/10.1007/978-3-319-91803-7_42. Last Accessed on 8 Oct 2023
3. 2019. 62 million in India visually impaired—thehindu.com. https://www.thehindu.com/news/cities/chennai/62-million-in-india-visually-impaired/article27699902.ece. Last Accessed on 8 Oct 2023
4. 2020. A new keyboard for typing braille on Android—blog.google. https://blog.google/products/android/braille-keyboard/. Last Accessed on 8 Oct 2023
5. 2022. Thinkerbell Labs and braille literacy in India—medicircle.in. https://medicircle.in/thinkerbell-labs-and-braille-literacy-in-india. Last Accessed on 8 Oct 2023
6. 2023. Vision impairment and blindness—who.int. https://www.who.int/news-room/fact-sheets/detail/blindness-and-visual-impairment. Last Accessed on 8 Oct 2023
7. Aksentseva A (2023) Mobile accessibility testing guide for product owners—mobidev.biz. https://mobidev.biz/blog/mobile-accessibility-testing-explained. Last Accessed on 8 Oct 2023
8. Alnfiai M, Sampalli S (2018) Improved singeltapbraille: developing a single tap text entry method based on grade 1 and 2 Braille encoding. J Ubiquitous Syst Pervasive Networks 9(1):23–31
9. Bonner MN, Brudvik JT, Abowd GD, Keith Edwards W (2010) No-look notes: accessible eyes-free multi-touch text entry. In: Pervasive computing: 8th international conference, Pervasive 2010, Helsinki, Finland, 17–20 May 2010. Proceedings 8. Springer, Berlin, pp 409–426
10. Burton D (2007) Talk me through it: a review of two cell phone-based screen readers—afb.org. https://www.afb.org/aw/8/1/14407. Last Accessed on 8 Oct 2023
11. Brian F, Southern C, Romero M (2011) Brailletouch: mobile texting for the visually impaired. In: Universal access in human-computer interaction. Context diversity: 6th international conference (UAHCI 2011), Held as Part of HCI International 2011, Orlando, FL, USA, 9-14 July 2011, Proceedings, Part III 6. Springer, Berlin, pp 19–25
12. Green N, Kruger J, Faldu C, St. Amant R (2004) A reduced QWERTY keyboard for mobile text entry. In: CHI'04 extended abstracts on human factors in computing systems, pp 1429–1432
13. Guerreiro T, Nicolau H, Jorge J, Gonçalves D (2009) NavTap: a long term study with excluded blind users. In: Proceedings of the 11th international ACM SIGACCESS conference on computers and accessibility, pp 99–106
14. Hamzah R, Mohamad Fadzil MI (2016) Voice4Blind: the talking braille keyboard to assist the visual impaired users in text messaging. In: 2016 4th International conference on user science and engineering (i-USEr). IEEE, pp 265–270
15. Jadhav A, Patil A (2012) Android speech to text converter for SMS application. IOSR J Eng 2(3):420–423

16. Jadhav D, Bhutkar G, Mehta V (2013) Usability evaluation of messenger applications for android phones using cognitive walkthrough. In: Proceedings of the 11th Asia Pacific conference on computer human interaction, pp 9–18
17. Naveen Kumar M, Linga Chandar PC, Venkatesh Prasad A, Sumangali K (2016) Android based educational Chatbot for visually impaired people. In: 2016 IEEE International conference on computational intelligence and computing research (ICCIC). IEEE, pp 1–4
18. Landicho JA (2016) VOISEE COMMUNICATOR: an android mobile application for hearing-impaired and blind communications. Int J Interact Mob Technol 10(4):26–31
19. Misra S, Banubakode SM, Dhawale CA (2014) Novel user interface for text entry on touch screen mobile device for visually impaired users. In: 2014 Global summit on computer & information technology (GSCIT)
20. United Nations (2023) Population|United Nations—un.org. https://www.un.org/en/global-issues/population. Last Accessed on 8 Oct 2023
21. Nayak S, Chandrakala CB (2020) Assistive mobile application for visually impaired people
22. Nour Azimah Z, Abdullah JM (2020) The Malaysian Journal of Medical Sciences' performance report for 2019. Malays J Med Sci 27(5)
23. Rodrigues A, Nicolau H, Montague K, Guerreiro J, Guerreiro T (2020) Open challenges of blind people using smartphones. Int J Hum Comput Interact 36(17):1605–1622
24. Romero M, Frey B, Southern C, Abowd GD (2011) BrailleTouch: designing a mobile eyes-free soft keyboard. In: Proceedings of the 13th International conference on human computer interaction with mobile devices and services, pp 707–709
25. Sagale U, Bhutkar G, Karad M, Jathar N (2018) An eye-free android application for visually impaired users. In: Ergonomics in caring for people: proceedings of the international conference on humanizing work and work environment 2015. Springer, Berlin, pp 291–297
26. Sánchez J, Aguayo F (2007) Mobile messenger for the blind. In: Universal access in ambient intelligence environments: 9th ERCIM workshop on user interfaces for all, Königswinter, Germany, 27–28 Sept 2006. Revised Papers. Springer, Berlin, pp 369–385
27. Senjam SS, Manna S, Bascaran C (2021) Smartphones-based assistive technology: accessibility features and apps for people with visual impairment, and its usage, challenges, and usability testing. Clin Optom 2021:311–322
28. Wang J, Zhai S, Canny J (2010) SHRIMP: solving collision and out of vocabulary problems in mobile predictive input with motion gesture. In: Proceedings of the SIGCHI Conference on human factors in computing systems, pp 15–24
29. Zhang MR, Wang R, Xu X, Li Q, Sharif A, Wobbrock JO (2021) Voicemoji: emoji entry using voice for visually impaired people. In: Proceedings of the 2021 CHI conference on human factors in computing systems, pp 1–18

Navigational Integration: Enhancing User Experience Through Illustration-Driven Wayfinding in a Multi-storied Building in Real and Augmented Environments

R. Gokul and Ankita Roy

Abstract Navigating a multi-storied building without a proper signage system can overwhelm and frustrate visitors, leading to a poor user experience. Finding specific offices, restrooms, elevators, and emergency exits can be challenging and intimidating, especially for first-time visitors. This can result in worry, delay in decision making, dilemma, leading to a negative user experience. In this paper we have described the results of an experimental study done to understand the effectiveness of an illustration based indoor navigation experience in real and augmented environments of a newly constructed building without any signage system in place. The main goal of this research was to understand how illustration in a building can enhance way-finding and effectively enable a user to navigate in a space. One approach to this experiment is to integrate navigation systems with colors, icons, forms and illustrations. This experiment further included visual aids, such as floor maps and directional signs, informational signs, statutory warning signs, etc. to enable a better understanding of the building's layout and easy accessibility to the intended destinations. This experiment was further validated with user studies and survey data from 35 participants which were analyzed along with interviews and observational empirical studies. The Second approach to the problem was dealt using Augmented Reality (AR) techniques with indoor space mapping and recreating the entire real world scenario to test the effectiveness before final implementation. In conclusion, the results suggest that incorporating graphics into navigation systems can significantly improve the user experience in multi-storied buildings, along with usage

R. Gokul (✉) · A. Roy
Indian Institute of Technology Hyderabad, IITH Road, Near NH-65, Sangareddy, Kandi, Hyderabad, Telangana 502285, India
e-mail: iamgokulrajenan@gmail.com

A. Joshi and G. R. Sim (eds.), *Proceedings of the 14th Indian Conference on Human-Computer Interaction*, Studies in Computational Intelligence 1162,
https://doi.org/10.1007/978-981-97-4335-3_8

of technological driven augmented environments. However, there may be challenges when merging large scale navigation systems with illustrations along with traditional signage systems, due to cost feasibility and other issues. Visitors can navigate the building more quickly, effectively, and confidently by providing precise and understandable directions. Careful planning and addressing integration issues can create a more user-friendly environment for all visitors.

Keywords Illustration · User experience · Signage system · Wayfinding · Interactive · Navigation · Interior · Multi-storied building · Augmented reality navigation · Usability test

subclass• **Human-centered computing → Empirical studies in HCI**

1 Introduction

This research is an experimental study done for a newly constructed building in an educational institution in Hyderabad, India, where the main requirement was to address the needs and issues of wayfinding and proper navigation. Since the entire building space is designed for a purpose, our main focus was to understand the key usage of the space and how design interventions can create effective solutions. The study began with a systematic desk research followed by actual site visits and conducting empirical research of the space. To understand the space further detailed interviews were conducted with various stakeholders. The authors also had the opportunity to interact with the startups during Innovation Day, where they showcased their ideas to potential investors and supporters. They discovered startups across various domains, such as automobile, software, blockchain, waste management, and electric innovations. Through our research and user studies, it was found that there is an innate need for vibrant spaces which will connect with the people who use the space. The research further investigates the effectiveness of an illustration based indoor navigation experience in real and augmented environments. The use of illustration in navigation systems offers several advantages. Firstly, it greatly enhances the user experience by reducing uncertainty and providing clear guidance, making visitors feel more at ease and less anxious. Secondly, it improves efficiency as visitors can locate their destinations more quickly and with fewer interruptions. Additionally, interactions with other stakeholders of the building revealed their vision for the space, aiming for a welcoming and encouraging yet professional space with a functional design.

2 Literature Review

2.1 Understanding Navigation and Wayfinding

Navigation is coordinated and goal-directed movement through the environment by organisms or intelligent machines. It involves both planning and execution of movements. It may be understood to include the two components of locomotion and wayfinding [20]. The word "navigation" originates from Latin and is derived from the words navis meaning "ship" and agere meaning "to act" [12]. Our main focus in this research is to understand navigation with wayfinding. Wayfinding is planning and decision making coordinated to the distal as well as local surrounds [20]. Research reveals that Humans and animals use various sensory elements and cognitive systems to navigate in a space and form a mental model to repeat the same. Humans also use symbolic representations to maintain orientation, including language and cartographic maps [20]. And the factors related to human psychology, such as orientation, attention, and the automaticity of wayfinding tasks; factors related to the environment, such as differentiation, visual access, layout complexity, and sign placement; and factors related to technology, such as the design of information displays [13]. This theory further intrigued us to further investigate the effectiveness of illustration based wayfinding for navigation. To effectively navigate in an environment, people resort to three distinct types of spatial knowledge: landmark knowledge, route knowledge, and survey knowledge [21]. The cognitive processes involved in making individual decisions include perception, attention, memory, and decision-making [16, 20]. Further to narrow down, navigation within a closed structure was investigated.

2.2 Understanding the Space

This research was conducted for a newly constructed multi-storied building in an institutional setup in Hyderabad which is designed to nourish and support the budding entrepreneurs. The building offers pre-incubation and incubation programs to assist entrepreneurs in creating prototypes and turning them into marketable products. This program aims to support start-ups in converting their prototypes into sellable products and gaining traction.

Business incubators support entrepreneurs in developing their businesses, especially in the initial stages. These are organizations geared towards speeding up the growth and success of start-ups and early-stage companies [15]. Start-ups in India are not recent; the start-up culture started in the '80s. We can see IT pioneers such as TCS, Wipro and Infosys leveraging technology and young people in building up the industry. The result is an industry that employs nearly four million people and generates 150 bn dollar in revenues[7]. Nearly 40 The combination of progressive government policies, capital assistance from corporates, guidance from academics and the sheer brilliance of Indian entrepreneurs cumulatively can create the secret sauce required to unleash the true potential of our start-up ecosystem [7]. While understanding the building structure, we came to a conclusion that it follows a brutalist architectural style characterized by exposed raw concrete and utilitarian designs. The sheer exposure of original materials such as steel, bricks and ferro-concrete makes the space unique. Brutalism emerged in the mid-twentieth century and is often associated with schools, churches, libraries, theaters, and social housing projects; Brutalism is often intertwined with twentieth century urban theory that looked toward socialist ideals [23]. As the monotonous outlook makes the space dull and due to which impacts the people within the space, while bringing in more visually appealing elements helps the people who inhabit the space for different purposes in many ways.

2.3 The Significance of Illustrations: Enhancing Spaces and Stimulating Minds with State of the Art Technologies

Illustrations enhance environments by stimulating creativity and fostering a sense of community. They bring vibrancy to spaces through mural art and wall illustrations, with a surge in environmental art beautifying urban centers and homes. Illustrations go beyond filling blank spaces, conveying stories and capturing the human condition. Furthermore, illustrations offer numerous benefits. They boost productivity, inspire creative thinking, leave a lasting impact on viewers, set the ambience of a space, and facilitate identification and orientation. The human brain processes images 60,000 times faster than text, and 90% of information transmitted to the brain is visual [14]. The influence of illustrations extends across various domains, transitioning from utilitarian purposes to an experiential realm. They transform the world into a more colorful and imaginative reality, surpassing the traditional medium of magazines. Virtual graphics, metaverses, and digital spaces have become new platforms for illustrations to flourish. Metaverse is already being used to enhance user experience and provide authentic brand communication strategies. Artificial intelligence enables companies to have high levels of targeted campaigns on digital platforms. Also usage of Augmented reality enriches users journey and creates an impactful experience along with extending the usual real world experiences [18].

2.4 The Impact of Illustration on Office Employees

Illustrations within an office space can have various effects on employees, categorized as follows:

Impact on Productivity: Research has shown that illustrations inside an office can positively impact employee productivity. If offices are to be designed for maximum productivity, then the dynamic nature of interaction needs to be integrated into the design of office environments [8]. The workplace environment is the most critical factor in satisfying an item in today's business world [1].

Impact on Well-being: Illustrations can reduce stress and anxiety, increase job satisfaction, and enhance creativity. An environment with art and illustrations reports higher well-being and job satisfaction. A functionally uncomfortable environment causes workspace stress and reduces the Quality of Work Life. [24] Architectonic details, which include colours and decoration, signage, artwork and design details, convey meaning and can have symbolic significance that affects people emotionally [17].

Impact on Creativity: That exposure to art can inspire creativity and increase the likelihood of employees coming up with new ideas. Numerous studies have demonstrated the positive effects on worker morale and productivity of occupying a workspace where ambient conditions and aesthetics are appropriate [24]. Art connected to the organization's mission, rotating exhibitions and diverse collections are valued by workplace viewers [22].

2.5 *Environmental Graphics: Enhancing Spaces and Navigating Experiences*

Environmental graphic design (EGD) combines various disciplines to enhance the physical environment visually, improving user experiences and creating emotional connections. EGD encompasses signage, identification, super graphics, art installations, and colour usage. It goes beyond basic wayfinding to create a layered and immersive spatial experience, incorporating elements of architecture, landscape, lighting, and more. EGD aims to connect people to places, fostering mental, physical, emotional, and spiritual connections. EGD enables individuals to form deep relationships with locations by translating physical features into stories. EGD is not only a decorative element but also has an important role as a medium of information, both as a directional sign and as an identity for a room [9]. These connections are essential for fulfilling human needs and enhancing a place's overall experience and memory. Interpretation helps people recognize a site by telling information about the site. While placemaking signifies a site through images or other identifying marks as a distinguishing feature of a site. These three components are integrated and support each other's information comprehensively to guide people in a site [9] which are crucial components alongside landmarks that assist in wayfinding and evoke a sense of familiarity for frequent users [24]. EGD and Wayfinding refer to information systems that guide people through a physical environment and enhance their understanding and experience of the space [5]. The three main components of EGD consist of signage, placemaking, and interpretation. Signage and wayfinding directing and navigating people to the destination site. In Cyprus, a case students were able oriented themselves within the building when the EGD elements were added to the building, as in the beginning everything was blank and didn't have a sense of belonging to them and not to navigate because of the insufficient signages [2].

2.6 Augmented Illustrations

Augmented Reality (AR) techniques were used with indoor space mapping and recreating the entire real world scenario to test the effectiveness before final implementation. Also adding AR guidelines creates multiple awareness levels for the visitors and creates additional extended experience of the space. Giving ways for exploratory easy navigation and wayfinding strategies by minimizing the time and walking distance. Recently, with the increasing popularity of AR functionalities integrated into mobile devices, researchers devised mobile authoring tools for creating and telling AR stories. For example, Rumiński et al. [19] created a mobile AR authoring tool named MARAT [11]. Chen et al. proposed SceneAR [4], which was a mobile application for creating sequential scene-based AR narratives.

3 Research Methods

Three steps of investigation were done to verify the fictitious pain points. The research was initially conducted through a survey to determine whether the illustrations impacted a multi-story institutional building. Additional interviews were conducted to learn more about the attitudes of those who would use the space. To validate the basic navigational issues, an experiment was finally carried out to grasp them. An empirical observational study was also conducted to see the users movement, decision making issues and other navigational dilemmas.

3.1 Stakeholders of the Project

Stakeholders in the project have been divided into three groups: primary, secondary, and tertiary. Primary stakeholders, including professors, startups, and staff members, who have full access and spend significant time in the area. Secondary stakeholders, such as students and interns, occupy the space for shorter periods but have the potential for a more prolonged engagement. Tertiary stakeholders, represented by visitors, have limited access and become involved at the authority's discretion. Recognising the importance of these stakeholder grouping them is crucial for this research.

3.2 Hypothetical Pain Points

As per the empirical observational study, the research pointed towards a few hypotheses that space lacks visual variety and has a chaotic navigation system, leading to a negative user experience. It hampers brand promotion and marketing efforts, while multiple doors add to the chaotic environment and hinders collaboration and positive work culture.

3.3 *Survey and Interview Questions*

To carry out the further research and validate the necessity and importance of the illustrations within the building, survey and interview was conducted. The questions are as mentioned in below sections:

Survey Questions: We are interested in your perspective on the role and impact of illustrations in multi-storied buildings. Firstly, do you believe that illustrations serve a meaningful purpose in such settings, and do you find them appropriate? Secondly, would you like to see an increased presence of illustrations in multi-storied buildings in the future? Additionally, we would like to gauge the significance of illustrations in enhancing the overall user experience within multi-storied office spaces. Have you ever visited a multi-storied building where illustrations and environmental graphics were utilized? If so, could you describe the type of illustrations employed and share whether they made the building appear more inviting and welcoming? Did these illustrations contribute to a sense of organization and structure within the building? Furthermore, did they make it easier for you to navigate and remember your way around the premises? On a scale of 1–5, how would you rate the ease of navigation in a multi-storied building that incorporates illustrations, and how much more enjoyable is the overall experience in such a building? Have you ever recommended a multi-storied building to others specifically because of its design and use of illustrations, and how visually appealing do you find these illustrations? Lastly, apart from illustrations, have you encountered any alternative methods to enhance the user experience within multi-storied office spaces? If so, we would appreciate your insights into these alternative approaches.

Interview Questions: In considering your perception of a building adorned with illustrations, we would like to gather your insights on several aspects. Firstly, what impression do these illustrations create for you? Additionally, in terms of style, what types do you believe would be most fitting to complement the building's aesthetics or which specific style do you think is most appropriate for this space? Furthermore, do you have a preference for any particular color that you believe should be prominently featured in these illustrations? Is there any specific theme or topic that you feel should be incorporated into these illustrations? We are also interested in whether you have encountered similar illustrations in interior spaces before and how you generally feel about them from a visual perspective. Moving forward, we would like to explore the impact of these illustrations on your overall perception of the building. Do you find them visually appealing, and do you think they enhance the user experience within the building? Specifically, do you believe these illustrations aid in improving wayfinding within the building, and do they contribute positively to the overall design? If possible, could you provide specific instances of how these illustrations have assisted you in navigating through the building? Beyond the realm of illustrations, we are curious to learn if you have any suggestions on additional measures that could further enhance the user experience of the building. Lastly, are there any other design elements within the building that you believe have played a role in shaping your overall experience? Your insights into these aspects are highly valued in our evaluation of the building's design.

3.4 *Insights from Survey*

To understand the relevance of illustrations in a workspace, the survey insights were analyzed. The purpose was to study the larger cultural behaviors associated with the topic, conduct user surveys to collect quantitative data and experiment with potential users to gather meaningful insights into their perspectives, concerns and motivations. The survey had 35 responses, with most users acknowledging that the illustrations enhance the space more visually appealing and increase productivity. Users have encountered a variety of illustrations, including ones that are highly decorative, abstract, minimal, and realistic. Additionally, according to the survey results, the illustration simplified navigating and enhanced the user's enjoyment of the space. Following are some of the key conclusions (Fig. 1).

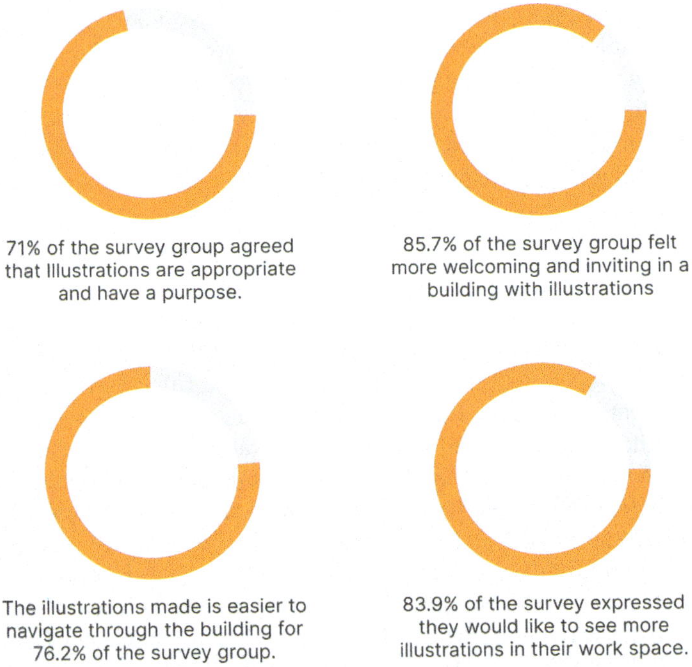

Fig. 1 The survey results showcasing the purpose and effect of illustration in a multi storied building for navigation and wayfinding

3.5 *Insights from Interviews*

The interviews mainly focused on understanding the mindset of different people who would be part of the space through long discussions and gathering qualitative responses. Another perspective was also kept in consideration that this would impact the next generation who would be becoming part of this entity. The keys insights were the following.

The illustrations should be minimal, simple, and modern. It should be a functional design rather than just a visually appealing design. The artworks should be related to the genres that they are working on. The elements related to nature make the space merge with the surroundings. People also stated the space to be lively and flavorus and accounted for the aspects of innovation. They further suggested the usage of latest state of the art technological interventions to create more valuable and memorable experiences. More elements point towards green initiatives and the ecosystem. Adding the factors which support the techy aspect, also details of Hyderabad's unique modern culture.

3.6 *Space Analysis and Identification of User Problems*

During a thorough empirical study and space analysis, potential issues were observed from the perspective of first-time visitors. The observational study was conducted without giving the participants any navigational clue. Further this experiment conducted with these participants resulted in confirmed hypothetical problems, prompting a physical assessment to validate pain points and identify additional issues.
Insights from the Experiment. The findings suggested that there were navigational issues in the building. Since the doors were placed in front of the stairs, even people were having trouble finding the staircases. The building had four stairs, but most people only noticed up to three of them. Additionally, they discovered that the rooms could not be distinguished from one another because everything appeared to be similar, but the size of the rooms varied. Although they had an easier time navigating because the layout was identical on the other floors, they still needed help to navigate. The experiment's key conclusions are as follows (Fig. 2).

4 Solutions

Several possible solutions were identified to improve user experience in a multi-storied building. These include implementing a wayfinding system using colours, patterns, or symbols, utilizing scannable illustrations for information, incorporating large-scale illustrations in key areas, developing a mobile app or gamified navigation system, and using LED sticker tags or AR apps. Physical navigation and illustrations

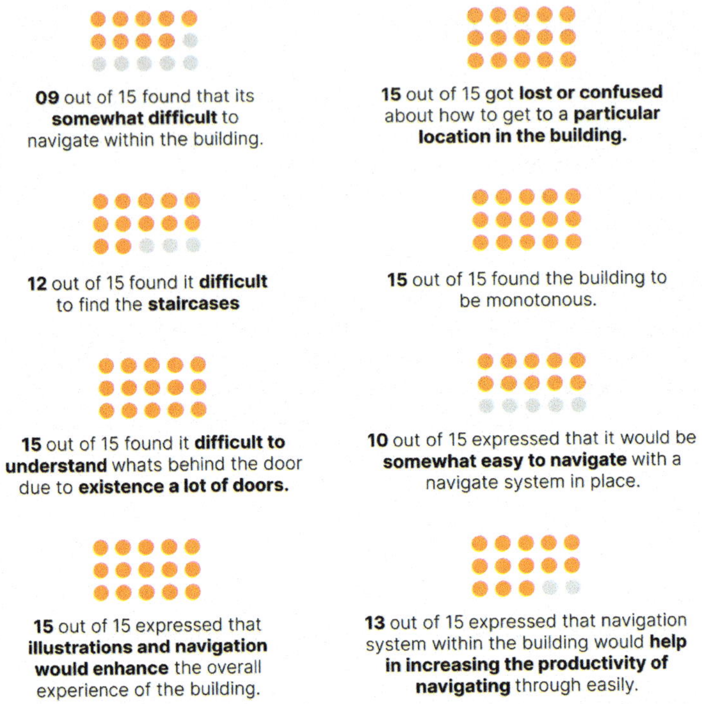

Fig. 2 The empirical study suggested navigational issues with qualitative feedback taken from every participant

were preferred due to cost-effectiveness, ease of use, and aiding in orientation and memorability. Card sorting with potential users helped determine relevant attributes for illustration. Further AR panels were created to understand the effectiveness of every illustration, to save cost before actual printable illustrated panels and for usability testing of the space.

4.1 Floor Map Analysis and Naming Segregated Sections for Improved Navigation

The floor map analysis divided the building into nine sections (A to I) to simplify navigation and reduce complexity. Pathways were identified using different colour codes, with most paths remaining consistent. Access to the building was limited to one point (Section A), and the navigation plan considered this limitation (Fig. 3).

Naming the sections with meaningful names was chosen to aid identification, create a personal connection, and facilitate effective communication. Generic labels like Section A, B, or C can confuse, while unique names simplify directions and meeting

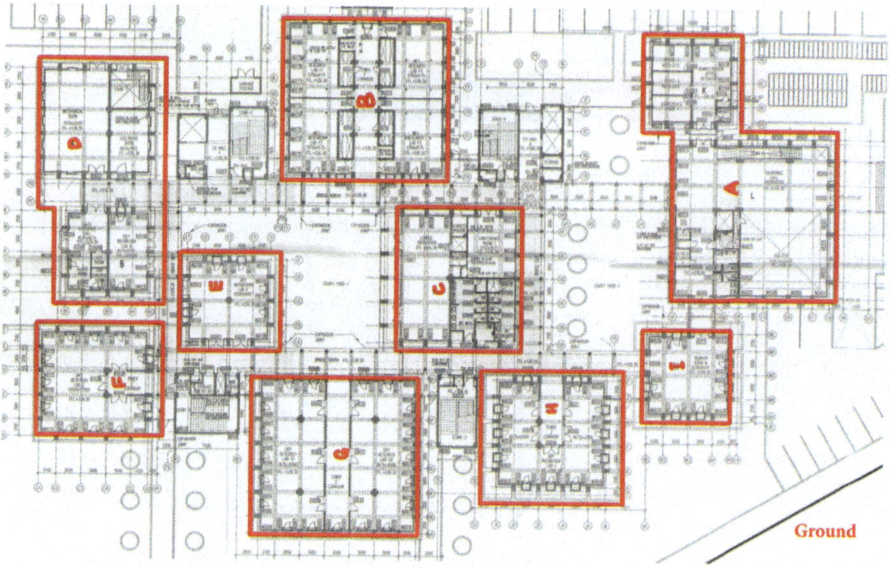

Fig. 3 The proposed building ground floor plan is divided into nine sections (suggestive floor map)

locations. The cognitive processes and related factors in indoor wayfinding in normal conditions have been extensively studied in prior research of Chen and Stanney, 1999; Darken and Peterson, 2001; Jul and Furnas, 1997 [3, 6, 10, 20]. Individuals make indoor wayfinding decisions through four cognitive processes, including perception, attention, memory, and decision-making, in normal conditions [3, 6, 10, 16, 20]

Initially, startup-related names were considered, which could lead to misinterpretation and confusion due to frequent changes in terms of incubation and space requirements. To ensure clarity and familiarity, names related to places, technology, and influential startup personalities were selected. The chosen names represent startup pioneers who revolutionized India, such as Ambani Avenue (Reliance Industries), Bansal Boulevard (Flipkart founders), Nilekani Nook (Infosys co-founder), Agarwal Alley (Ola founder), Kalra Boulevard (MakeMyTrip founders), Gupta Gallery (Urban Company founder), Vijay Grove (Paytm founder), Bansal Block (Snapdeal founders), and Siddhartha Plaza (Cafe Coffee Day founder). By dividing the space into distinct sections, implementing color-coded pathways, and assigning meaningful names, the navigation experience within the building is simplified and personalized. Users can easily identify and remember specific locations, fostering a sense of ownership, pride, and positive emotional connection with the surroundings. Effective communication is facilitated, as users can navigate using familiar names and landmarks, reducing ambiguity and enhancing overall navigation efficiency.

Fig. 4 Colour-coded navigation arrows with section names

Fig. 5 Simple ground plan represents how the color coded navigation is placed on the floor

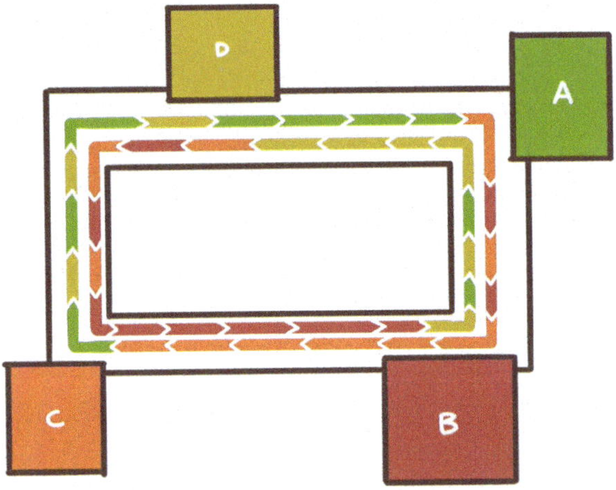

4.2 *Final Navigation Idea: Colour-Coded Arrows for Easy Building Navigation*

The final navigation idea incorporated a colour-coded system with two lines, one clockwise and the other anti-clockwise, to enhance navigation within the building. This approach simplifies the ideation process and offers a straightforward way for users to move through the space (Fig. 4).

The colour-coded arrows at each entry point serve as directional guides for users, eliminating the need to refer to descriptive boards. The arrows change colour according to the destination, ensuring straightforward and intuitive navigation. The shortest distance is considered to optimize efficiency, and the same colour does not appear in both lines at the exact location (Fig. 5).

This visual representation demonstrates how the colour-coded arrows aid users in navigating between sections. Since the model closely resembles the existing building, the implementation should be seamless and hassle-free. With this final navigation idea, users can quickly glance at the arrows and take appropriate actions, facilitating a smooth and intuitive navigation experience throughout the building.

4.3 *Motivational Theme and Color Palette for Illustrations*

Motivational Theme: A motivational theme has been chosen to create an inspiring atmosphere within the building. It promotes productivity, creativity, and a positive outlook. Incorporating inspirational illustrations is a constant source of strength and motivation for individuals, encouraging them to overcome obstacles and embrace the entrepreneurial spirit. This theme fosters community and enthusiasm among startups in the building.

Colour Palette: The colour palette for illustrations was considered in two ways. Common spaces like the cafeteria will feature lively and varied colours to create an energetic ambience. On the other hand, illustrations supporting the navigation system will have a monochromatic approach with accent colours. This ensures visual coherence and clarity while guiding users through the building, maintaining a consistent aesthetic.

5 Explorations

Examining illustrations in the field reveals a varied landscape with different methods and styles. Understandably, the project at hand seeks to produce semi-realistic pictures with defined elements and a simple yet effective perspective. The semi-realistic approach makes it possible to balance capturing recognisable details and adding a dash of whimsy and fun. This strategy resonates with a wider audience because it combines familiarity with imagination.

The intentional use of minimal and simple perspective further enhances the sense of place within the illustrations. This method ensures that viewers can understand and easily explore the areas being portrayed. By eliminating unnecessary complexity, it makes a more intuitive and user-friendly experience.

6 Final Illustration Designs and Functionality

Illustrations for the Common Spaces: Vibrant illustrations with a colourful palette are used in common areas to create a lively and engaging environment, adding vibrancy to the overall ambience (Fig. 6).

Illustrations on Doors: Illustrations on doors help users understand what is behind them, providing valuable information and guiding them to take necessary precautions (Fig. 7).

Supporting Illustrations for Navigation System: Monochromatic illustrations with a few accent colours support the colour-coded navigation system, aiding users in finding their way through the building (Figs. 8 and 9).

Fig. 6 Illustration for the cafeteria and common areas

Fig. 7 Illustration for the doors

Fig. 8 Supporting illustrations for the navigation system

Fig. 9 Mock-ups of the supporting illustrations for the navigation system

Fig. 10 The user successfully navigates to their destination by consulting the lobby's navigation board, following the colour-coded line, and recognising the corresponding illustration

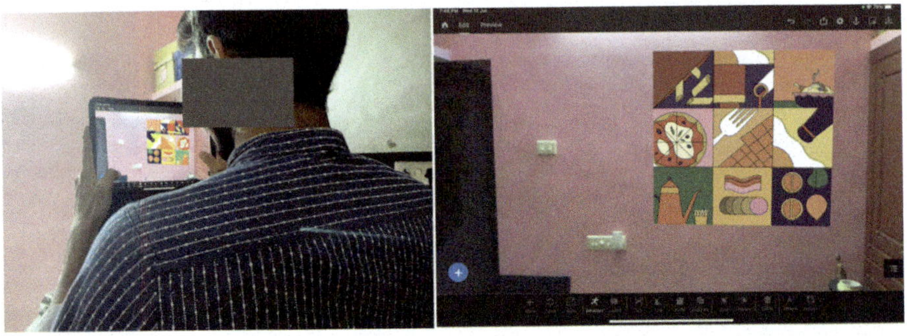

Fig. 11 Testing the functionality of the AR application

7 How Does the System Work?

Users refer to a navigation board placed in the building's lobby or entrance to determine their destination. They analyse the names and colour codes provided on the board. Then, they follow the corresponding colour-coded lines on the floor, designed to guide them efficiently to their destination. Along the way, users encounter illustrations that validate they are on the right track, boosting their confidence and enhancing their sense of location, productivity, and creativity. The illustrations' visual cues spark users' curiosity and thought, making their journey more engaging and enjoyable (Fig. 10).

8 AR Application for Illustrations

These digitally created images, like those used in AR applications, can be utilized to determine whether something fits the intended environment. To reduce the likelihood of implementation errors, the programme will let you understand the essential adjustments that can be made even before performing the examples in the relevant regions (Fig. 11).

Fig. 12 Scannable illustration

A bigger audience can obtain information on what happens behind the scenes, even if these people are not physically present in the area, by including elements in the illustrations that can be scanned, such as QR codes. Recently, some AI-produced scannable images, making it increasingly clearer how this concept can be developed (Fig. 12).

9 Usability Testing

Augmented Reality was chosen as one of the tools to test the Workability of the entire environmental graphics and navigational wayfinding efficiency. To understand it further the illustration and wayfinding holistic solutions were incorporated in the AR system through the software Adobe Aero, where screens were developed for every section and the real time image of the actual space was tracked and addition of virtual environmental graphics were added. This also included the floor graphics and assistive information to navigate the users. The entire system was then shown to participants (stakeholders) for usability testing to visualize the actual implementation scenarios and understand the cost implications. Specifically, the testing sought to determine whether these elements facilitated a more intuitive and efficient navigation process within the building. It allowed stakeholders to engage with the design concepts in a realistic setting, providing valuable insights into the user experience.

In summary, the usability testing of illustrations and color-coded arrows within the building's AR application proved to be a sophisticated and insightful approach. It effectively gauged the usability of these design elements, ensuring their potential benefits in enhancing navigation while avoiding the pitfalls of premature implementation.

10 Conclusion

Integrating motivational illustrations and a colour-coded navigational system has transformed the building into an inspiring, user-friendly space. The motivational theme, conveyed through visuals and messages, fosters a positive mindset and motivates occupants in their entrepreneurial endeavours. The clockwise and anticlockwise navigation system ensures seamless wayfinding. Prospects include expanding the motivational theme with sub-themes and incorporating technologies like augmented reality for interactive experiences. This collaborative approach creates a shared inspiration and supports the occupants' entrepreneurial journey. The project sets the stage for an evolving and dynamic environment within the startup incubation centre, empowering occupants and promoting their success.

References

1. Ajala EM (2012) The influence of workplace environment on workers' welfare, performance and productivity. Afr Sympos 12:141–149
2. Atamaz E (2019) Environmental graphic design for building information systems. EurAsian J BioSci 13(2)
3. Chen JL, Stanney KM (1999) A theoretical model of wayfinding in virtual environments: Proposed strategies for navigational aiding. Presence 8(6):671–685
4. Chen M, Monroy-Hernández A, Sra M (2021) SceneAR: scene-based micro narratives for sharing and remixing in augmented reality. In: 2021 IEEE International symposium on mixed and augmented reality (ISMAR). IEEE, pp 294–303
5. Clark C (2018) What is wayfinding? Part 1: it's all about human needs. Retrieved 08 April 2023 from https://rsmdesign.com/news/what-is-wayfinding-part-1-its-all-about-human-needs
6. Darken RP, Peterson B (2014) Spatial orientation, wayfinding, and representation
7. Gulshan YER (2023) India's startup story, unlocking true potential. Retrieved April 08, 2023 from http://media-publications.bcg.com/bcg-sift-1-3-indias-startup-story-unlocking-true-potential.pdf
8. Haynes BP (2008) Impact of workplace connectivity on office productivity. J Corp Real Estate 10(4):286–302

9. Indriati L, Setiawan JL, Engel MM (2022) Embracing the future: creative industries for environment and advanced society 5.0 in a post-pandemic era
10. Jul S, Furnas GW (1997) Navigation in electronic worlds: a CHI 97 workshop. ACM SIGCHI Bull 29(4):44–49
11. Li W, Li C, Kim M, Huang H, Yu L-F (2023) Location-aware adaptation of augmented reality narratives. In: Proceedings of the 2023 CHI conference on human factors in computing systems, pp 1–15
12. Luzum B (2004) Navigation principles of positioning and guidance
13. Montello DR, Sas C (2006) Human factors of wayfinding in navigation
14. I. F. o. V. Practitioners (2022) This is why our brain loves pictures. Retrieved 11 Sept 2023 from https://ifvp.org/content/why-our-brain-loves-pictures
15. Government of India (2023) Business incubators. Retrieved 11 Sept 2023 from https://aim.gov.in/aic-overview.php
16. Orquin JL, Loose SM (2013) Attention and choice: a review on eye movements in decision making. Acta Psychologica 144(1):190–206
17. Richter K-F, Winter S (2014) Introduction: what landmarks are, and why they are important. In: Landmarks: giscience for intelligent services, pp 1–25
18. Rizvanović B, Zutshi A, Grilo A, Nodehi T (2023) Linking the potentials of extended digital marketing impact and start-up growth: developing a macro-dynamic framework of start-up growth drivers supported by digital marketing. Technol Forecast Social Change 186(2023):122128
19. Rumiński D, Walczak K (2013) Creation of interactive AR content on mobile devices. In: Business information systems Workshops: BIS 2013 international workshops, Poznań, Poland, 19–20 June 2013. Revised Papers 16. Springer, Berlin, 99 258–269
20. Shah P, Miyake A (2005) The Cambridge handbook of visuospatial thinking. Cambridge University Press, Cambridge
21. Siegel AW, White SH (1975) The development of spatial representations of large-scale environments. Adv Child Develop Behavior 10(1975):9–55
22. Smiraglia C (2014) Artworks at work: the impacts of workplace art. J Workplace Learn 26(5):284–295
23. Stewart J (2022) Brutalism: what is it and why is it making a comeback? Retrieved 12 April 2023 from https://mymodernmet.c/brutalist-architecture/
24. Vischer Jacqueline C, Mariam W (2017) The effect of workplace design on quality of life at work. In: Handbook of environmental psychology and quality of life research, pp 387–400

Arts Track Papers (Short Papers)

Belonging: Exploring Audio-Visual Cues in Real-Time 6DoF VR Narrative

Jayesh Pillai, Priyanshu Thakur, and Tushar Vaid

Abstract VR narratives provide viewers with the feeling of being immersed in a story environment by giving them the freedom to look around and change their point-of-view during the experience. Storytellers use various techniques to guide the viewers to important plot points in a VR narrative. Previous studies have examined various methods to design narratives for 360° VR films with linear and non-linear narratives. In this study, we look at the grammar of storytelling in a real-time 6 degrees of freedom (6DoF) VR experience. We created 'Belonging', a real-time VR narrative in which one has the freedom to experience the story by physically moving around in the environment and changing point-of-view as the story progresses. Through this project, we intend to explore design-related aspects of narrative and associated technical challenges for creating a compelling 6DoF real-time VR experience. In this narrative, we focus on the audio and visual cues that would effectively guide the viewers to the plot points.

Keywords VR film · Storytelling · 6DoF · Perceptual cues

CCS Concepts: • **Human-centered computing** → **Virtual reality**; • **Applied computing** → Media arts

---

J. Pillai
IDC School of Design, Indian Institute of Technology Bombay, Mumbai, India

P. Thakur
National Institute of Design, Vijayawada, India

T. Vaid (✉)
Next Tech Lab, SRM University, Chennai, India
e-mail: tusharvaid30@gmail.com

A. Joshi and G. R. Sim (eds.), *Proceedings of the 14th Indian Conference on Human-Computer Interaction*, Studies in Computational Intelligence 1162,
https://doi.org/10.1007/978-981-97-4335-3_9

1 Introduction and Related Work

While a traditional film narrative is presented using visuals that are bound to a rectangular frame, a VR film immerses the users in the story environment that appears to surround them. This provides the users with the freedom to choose where to look and to change their point-of-view (POV) constantly. This liberty to frame the visual content themselves brings about challenges for the storytellers in guiding the users so as to convey the intended narrative as effectively as possible [1]. Researchers and filmmakers exploring VR film are evaluating new storytelling methods to create efficient user experiences [2]. With the rising interest in creating and experiencing VR films, the traditional tools, techniques and principles are being reconsidered, leading to the definition of a new grammar of storytelling [1, 3].

Previous studies in this domain have explored this evolving grammar and have presented the significance of perceptual (audio and visual) cues and their impact on guiding the users' attention to different narrative elements (plot points) in a VR film [3–8]. The previous explorations were often associated with linear VR narratives. A linear (pre-rendered) VR film (such as in [3, 9]) contains strictly one narrative but with the freedom for the users to change their perspective in a panoramic 'look around' mode, which is referred to as a 3 degrees-of-freedom (3DoF) experience. A few studies on non-linear responsive VR films have also been conducted, such as Cinévoqué [9, 10], where the experience uses pre-rendered 360 videos but has been purely 3DoF experiences.

This project looks at potential storytelling techniques relevant to real-time VR films with six degrees of freedom (6DoF) experience, that allows a user to not just 'look around' but also 'move around' in the story environment. There have been a few remarkable projects on 6DoF VR films [11, 12] that helped us understand various storytelling techniques for a real-time VR Film. Here we look at the grammar of storytelling in VR that would aid in conveying the story effectively. Through this study, we intend to explore the design of a narrative for VR, particularly focusing on audio and visual cues.

2 Design of the Experience

What makes the storytelling different for a VR film different from a traditional film is that we must keep in mind where we want the viewer to be ideally located during a certain point in the story. This helped in shaping the story and the dialogues for the film.

'Belonging' is the story of the sculptures of a certain artist. On the night before the exhibition of the artist's latest works, the sculptures are arranged on their pedestals in an art gallery. When no one is around, they talk about themselves, their unique qualities, and their potential owners (Fig. 1). One among them is a sculpture who

Fig. 1 Sculptures conversing with each other in the exhibition space, in the VR narrative 'belonging'

struggles to understand her identity and her place in that world. Through the narrative, the viewer gets a glimpse into her mind as well as what she would at the end learn about her 'belonging'.

2.1 *The Narrative*

We developed the narrative to be set in the art gallery, where the viewer gets to be part of the conversations the sculptures have. The total space a viewer could move around and experience the story was designed to be 5 × 5 m. This space contains a corridor, a gallery space and a store room area. The narrative was designed to be a 6DoF room-scale experience where the viewer has to move around without having the option to teleport. In the gallery space, the arrangement of the sculptures and the order of the dialogues were designed in a manner that would prompt the viewer to move in a specific path.

2.2 *Perceptual Cues*

This study primarily focuses on how the various cues help the viewer navigate the space successfully in an intuitive way. The space and the timeline of the narrative were accordingly planned with careful attention to the timings of each plot point.

2.2.1 Space and Plot Points

There are three scenes in the narrative of 'Belonging'. The various plot points, along with the ideal path and direction of the viewer, are shown in Fig. 2, using a top-view of the story space. The placement and timing of these plot points were designed to guide the viewer gradually towards specific locations in the narrative space.

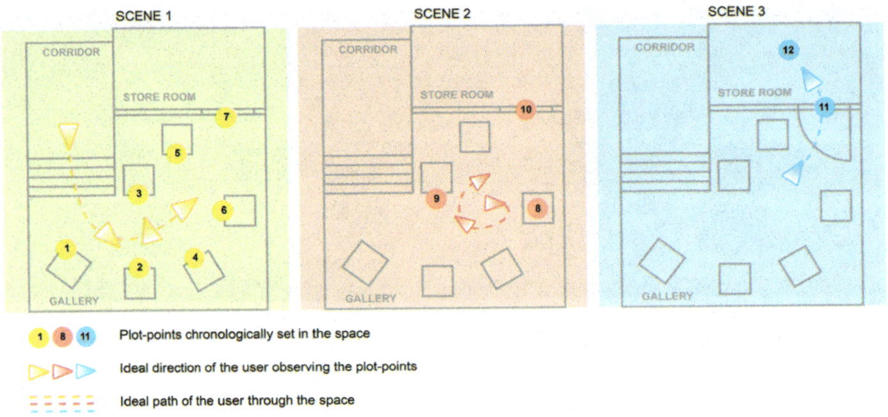

Fig. 2 Spatial representation of the narrative with plot points and ideal path of the viewer

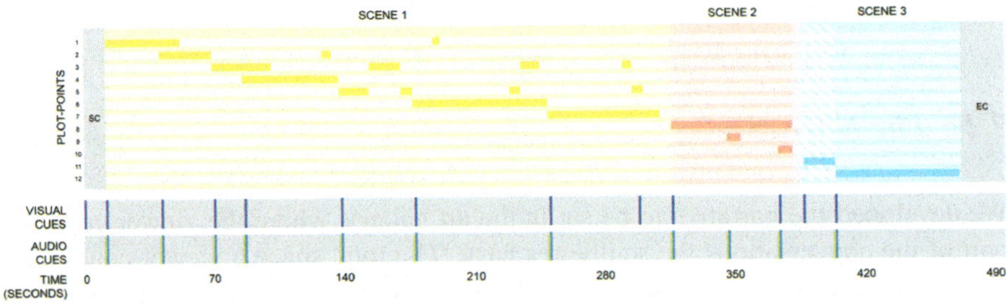

Fig. 3 Temporal representation of the narrative with audio and visual cues

2.2.2 Timeline and Audio-Visual Cues

It was essential to visualise the narrative in a temporal manner since there were multiple plot points taking place in parallel. The three scenes mentioned above, along with the starting and ending credits, are represented as a timeline in Fig. 3. The visual and audio cues that were designed to aid attention shifts from one plot point to another are also captured under the timeline in Fig. 3.

2.3 Technical Implementation

This VR narrative was created using Unreal Engine 5 (Fig. 4) with the 3D objects created in Blender. In order to overcome the technical challenges that came with the experience of real-time VR films, we used a few unique techniques that are explained below.

Fig. 4 Spotlights and spatial audio setup in the Unreal Engine 5 Editor

2.3.1 Triggering the Plot Points

In order to convey the story effectively, a real-time VR film may have plot points that are fixed as well as elements that may require to be triggered, depending on the position or the POV of the viewer in the story environment. Triggering plot points based on certain actions of the viewer helped in avoiding missing important aspects of the story while still keeping the narrative linear.

For instance, in the narrative (at plot point 11), the viewer is required to enter a room through a door where a certain event (plot point 12) is to happen. There are specific visual cues that inform the viewer to enter the room. A logic is attached to the door of the room that checks the distance between the viewer and the door, which if below a certain threshold, triggers the animations and dialogues.

2.3.2 Spatial Audio

Spatial Audio plays an important role in any VR film that relies on diegetic sounds [3, 13–15]. The dialogues, as well as the sound effects, use spatial audio, which is handled by Unreal Engine 5's attenuation system (Fig. 4). Due to this, the viewer is aware of which character is speaking from which direction and where the important events are happening. Sound effects like a door opening, spotlights turning on, and footsteps are used in this narrative that heavily relies on spatial audio to gain viewer attention.

2.3.3 Out of Bounds

One of the major challenges with room-scale VR experiences is helping the viewer stay within the desired location of the story without going out of the bounds of the environment. We overcame this by making the screen fade to black as the player walks towards the edge of the boundary. The screen will start fading in a linear fashion when the player is close to the wall and fade more as they get closer until it turns completely black on touching (or crossing) the wall. This was implemented by sending out four rays in left, right, forward and backward directions at the same time to hit the boundaries of the experience. The distance between the nearest wall is calculated, and once it is lower than a certain threshold, the fading process is triggered.

2.3.4 Natural Vertical Movement

Navigating the player in VR without creating motion sickness has been a challenge for the developers. Mechanics like walking up or down the stairs (or ladders or slopes) must be designed carefully so that they feel natural. In this film, we have added the functionality for the viewer to walk on steps and simulate an upward or downward motion as they move. We implemented this in the game engine by casting a ray vertically downwards from the camera to detect the ground plane below the viewer. It helps calculate the distance between the camera and the ground and accordingly moves the viewer up or down depending on whether the distance is lower or higher than the player's height. This makes physically moving around the environment much more immersive. Implementing and understanding such vertical movements in a VR space will help designers imagine narratives containing environments with multiple levels.

2.3.5 Character Facial Animations

The sculptures in the narrative have a distinctive form that required specific facial and body animations. In order to achieve a perfect sync for the lip animations with the dialogues, we implemented a unique method. The amplitude of the audio of the dialogue was used for moving the lips in real-time. This system made the testing of iterations of the dialogues remarkably fast during the prototyping stage, as it allowed for swapping audio files without having to change the entire animation. This was made possible by Unreal Engine 5's audio analyser plugin.

3 Prototype Testing

As a pilot study, our prototype was experienced by six participants. We observed the audio-visual cues that were designed for the narrative and examined if they worked effectively as planned. We found that the spatial audio cues were the most effective

Fig. 5 Participants experiencing the VR narrative

ones in guiding the viewer's attention. As the sculptures in the narrative remained in their positions while conversing with each other, there were no prominent movements of the characters from one point to another. This was also another reason the visual cues, in a few cases, became secondary to the audio cues. The design of the order in which the sculptures speak in the film proved to be helpful in the gradual movement of the viewer to the relevant plot points. Most importantly, we found that after the experience, the viewers were able to understand the essence of the story (Fig. 5).

4 Conclusion and Future Work

We believe that through this demonstration, we will be able to study how viewers experience stories in 6DoF VR narratives. Understanding audio-visual cues is key to understanding the evolving grammar of storytelling in VR. We intend to conduct an in-depth analysis of these perceptual cues and propose guidelines to help VR filmmakers in designing effective VR narratives.

References

1. Pillai JS, Ismail A, Charles HP (2017) Grammar of VR storytelling: visual cues. In: Virtual reality international conference 2017, Laval, France
2. Gödde M, Gabler F, Siegmund D, Braun A (2018) Cinematic narration in VR-rethinking film conventions for 360 degrees. In: Virtual, augmented and mixed reality: applications in health, cultural heritage, and industry: 10th international conference, VAMR 2018, held as part of HCI international 2018, proceedings, part II 10, Las Vegas, NV, 15–20 July 2018. Springer International Publishing, pp 184–201
3. Pillai JS, Verma M (2019) Grammar of VR storytelling: analysis of perceptual cues in VR cinema. In: The 16th ACM SIGGRAPH European conference on visual media production, London, UK
4. Pillai JS, Verma M (2019) Grammar of VR storytelling: narrative immersion and experiential fidelity in VR cinema. In: 17th ACM SIGGRAPH international conference on virtual reality continuum and its applications in industry (VRCAI), Brisbane, Australia

5. Wang H, O'Fearghail C, Zerman E, Braungart K, Smolic A, Knorr S (2021) Visual attention analysis and user guidance in cinematic VR film. In: International conference on 3D immersion (IC3D). IEEE, pp 1–8
6. Knorr S, Ozcinar C, Fearghail CO, Smolic A (2018) Director's cut: a combined dataset for visual attention analysis in cinematic VR content. In: Proceedings of the 15th ACM SIGGRAPH European conference on visual media production, pp 1–10
7. Nielsen LT, Møller MB, Hartmeyer SD, Ljung TC, Nilsson NC, Nordahl R, Serafin S (2016) Missing the point: an exploration of how to guide users' attention during cinematic virtual reality. In: Proceedings of the 22nd ACM conference on virtual reality software and technology, pp 229–232
8. He E, Lin J, Liu Z, Zhang Y (2017) Research on perceptual cues of interactive narrative in virtual reality. In: Human interface and the management of information. Information presentation and visualization, pp 283–296
9. Pillai JS, Murugan A, Dev A (2019) Cinévoqué: design of a passively responsive framework for seamless evolution of experiences in immersive live-action movies. In: 17th IFIP TC.13 international conference on human-computer interaction—INTERACT 2019, Paphos, Cyprus
10. Murugan A, Dev A, Pillai JS (2019) Cinévoqué: development of a passively responsive framework for seamless evolution of experiences in immersive live-action movies. In: The 25th ACM symposium on virtual reality software and technology, Sydney, Australia
11. Penney D, Chung E (2017) Arden's wake. In: ACM SIGGRAPH 2017 computer animation festival, p 28
12. Ustik G. Henry (or how VR will undermine the cinematic experience)
13. Bala P, Masu R, Nisi V, Nunes N (2018) Cue control: interactive sound spatialization for 360° videos. In: International conference on interactive digital storytelling. Springer, pp 333–337
14. Lescop L (2017) Narrative grammar in 360. In: IEEE international symposium on mixed and augmented reality (ISMAR-adjunct). IEEE, pp 254–257
15. Dowling D, Fearghail CO, Smolic A, Knorr S (2018) Faoladh: a case study in cinematic VR storytelling and production. In: International conference on interactive digital storytelling. Springer, pp 359–362

Whispers of the Konkan Guardian: Interactive Storytelling Through Digital Game Design

Shreyas Vernekar and Jayesh Pillai

Abstract "While walking down the road the other day, I heard a thumping noise and a rattling of ghoongroos ("Ghoongroo" is a Hindi word that refers to a musical accessory worn generally by dancers in traditional Indian dance forms. In English, it is commonly known as "ankle bells" or "jingles." These are small, metallic bells tied around the ankles of dancers that create a rhythmic sound when the dancer moves.). Instantly, I knew he was patrolling his village, and I dared not cross his path," the person recounted. "He is 10 feet tall and would squish anyone walking on his route with his long legs." Another individual shared a contrasting experience: "He helped me reach home. I thought he was someone from the village." Rakhandar, the mythical protective spirit of the area, demonstrates a fascinating duality in his identity. He can be mischievous, unforgiving, or fiercely protective towards the region's residents. This multi-faceted nature adds depth and intrigue to the legends surrounding Rakhandar, making him an enigmatic figure in local folklore. This project aims to bring the rich folklore and mythology of the Konkan region to a broader audience, focusing on the figure known by various names such as Rakhandar, Rakhnau, Naas, Rakhno, Devchar, Daivata, and more. These names represent a spirit god believed to consistently patrol villages at stipulated times, protect their inhabitants, and assist those in need. The project seeks to create an interactive and immersive environment that transports users to rural Goa, with abundant coconut trees, interesting stories,

Shreyas Vernekar and Jayesh Pillai contributed equally to this research.

S. Vernekar (✉) · J. Pillai
IDC School of Design, Indian Institute of Technology Bombay, Mumbai, India
e-mail: shreyasver.5@gmail.com

J. Pillai
e-mail: jay@iitb.ac.in

A. Joshi and G. R. Sim (eds.), *Proceedings of the 14th Indian Conference on Human-Computer Interaction*, Studies in Computational Intelligence 1162,
https://doi.org/10.1007/978-981-97-4335-3_10

and numerous Rakhandar shrines. Our intention is to preserve and celebrate the cultural heritage of the Konkan region while fostering a deeper understanding and appreciation of its mythology and traditions.

Keywords Game design · Interactive storytelling · Folklore experience · Immersive experience

CCS Concepts: • **Human-centered computing** • **Applied computing** → **Media arts**

1 Introduction

The concept of a protective spirit responsible for cultural, ecological and social preservation is widespread in and around Konkan Region. However, very little is known about the accompanying culture and spirit-god coexistence. Numerous stories are circulated about colossal anthropomorphic spirits that are consistently described across different sightings and encounters. These entities are commonly referred to by different names that vary across different regions and dialects. For the sake of consistency in this abstract, we will use the term "Rakhandar" as it is widely recognized and commonly used in mainstream contexts (Fig. 1).

The United Nations [8] highlighted the significance of preserving both physical and intangible aspects of culture. This focus is especially crucial in disadvantaged areas of the Global South, where cultural traditions and historical stories are steadily

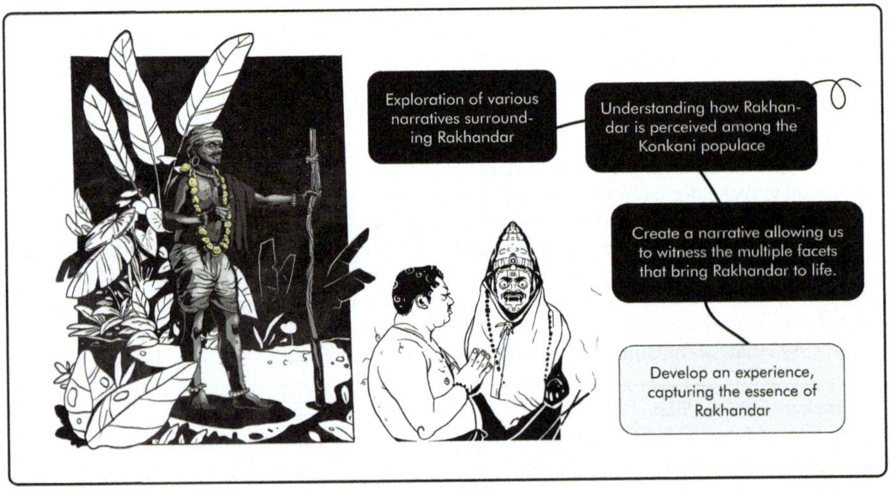

Fig. 1 Illustration of Rakhandar, based on photographs captured during Shigmotsav festival along with the project journey from ideation to execution

disappearing [1]. Given the significant presence of these stories and beliefs within the community influencing their day-to-day decisions, it is crucial to ensure that these stories are studied, and their significance understood in building a collective memory for the preservation of a community or ecological system.

In a world facing rapid population growth, migratory processes, and globalisation, there is an increasing interest in supporting intercultural education [6]. One effective way to foster intercultural understanding and learning is through the medium of video games. By creating an immersive virtual environment centred around Rakhandar culture, players can engage in an interactive and educational experience that transcends traditional boundaries. Within this virtual world, users will encounter moral dilemmas, learn about towering issues, participate in local offerings, and experience the significance of community dependencies.

2 Ideation

An immersive experience enables a user to empathize and increase involvement in the narrative. To give the user agency, opting for a non-linear open-world game could provide the most exposure to the virtually created rural Goa. It was important for the narrative to involve a relatable protagonist, which led us to create Sarvesh's persona, who does not belong to the village. This persona was crafted with the assumption that most users experiencing this interactive story have limited information about Rakhandar. Furthermore, we had to choose between a first-person perspective, a third-person perspective, and a completely VR experience for the gameplay style. Narratives presented in the third-person offer a broader view of the world beyond the limitations of a first-person account. On the other hand, employing a third-person limited perspective enables a deeper exploration of the character's actions, fostering a stronger sense of empathy for the reader [2]. Considering that we are aiming for a more story-driven narrative, allowing the user to make the character move or perform certain actions could help the user empathize with the character and thus make the story experience richer [7]. Hence, opting for a third-person perspective had a better advantage as opposed to the alternatives.

We designed a non-linear interactive story centred around Rakhandar culture, which then tapers into a linear narrative. The focal point of the narrative is the rituals, which involve offering traditionally made or collected items such as fish, alcohol, beedi, and local bread to the deity. These offerings spark engaging conversations between the main character and the villagers, considering the main character's previous exposure only to mainstream Brahmanic rituals. The main character collects all the necessary objects for the ritual, enriching his and, by extension, the player's understanding of Rakhandar. This segment of the story emphasizes Rakhandar's benevolent nature, highlighting its role as the protector and nurturer of the village.

To enhance player engagement and provide a more immersive experience, we included additional interactive elements within the game. These elements went beyond simple player conversations and offered engaging activities for the players to participate in. For example, Sarvesh interacts with environmental objects such as fallen mangoes and Feni-making distillery. Considerable effort was dedicated to delving into the intricate character dynamics and their deep-rooted connection with the village. The primary goal was to ensure substantial character growth, resulting in profound transformations by the conclusion of gameplay. This intentional focus on character development added a layer of depth, making the overall gaming experience remarkably engaging for players. For example, to provide Sarvesh, the city boy protagonist, with a guiding companion throughout his village journey, we introduced Pranav as a character with a contrasting personality. Pranav, a roughened village boy, was designed to complement Sarvesh's urban background and create humorous and engaging situations.

3 Research

3.1 Secondary Research

We have discovered a significant online presence of websites and blogs that serve as digital repositories for experiences and narratives related to this ancient tradition. Furthermore, the inclusion of audiovisual content, such as religious songs, horror stories, and narrated tales, has enriched our comprehension of how these stories are transmitted through spoken words. This multimedia content reflects the deep reverence and devotion directed towards Rakhandar, as evident through various interpretations within the community. We also discovered literary works dedicated to the exploration of supernatural entities in India. Particularly intriguing was the abundance of references specifically focused on Indian gods, ghosts, demons, and other mystical beings. Below are some excerpts which give us an insight into people's experiences and beliefs surrounding Rakhandar.

> Before leaving, we would fold our hands in reverence [...] and quietly walk away without talking. Thomas would warn me not to look behind. [5]

> People who break the law or destroy his land suffer the brunt of his fury. There are special traditions that are almost paganistic but ritualistically practised, avoiding disfavour with this legend. These traditions would differ from village to village. In some places *sur*, *kharem*, and *pao* are offered, while others might offer *rott* (rice roti) and, still others, the sacrifice of fowl. The place of offering would be borders of villages, bridges and fields. [3]

> In the Spring, during Shigmo, Betal temples hold a festival called Gadyachi Zatra. During this festival, in the middle of the night, Devchar is said to appear as a faceless, shadowy form. He delights the crowds by holding flames in his palms and walking in the air. Cameras are strictly forbidden at the festival. [4]

(a) Rakhandar shrine devoted to local Rakhandar, Shri Kunnekar Jogeshwar situated in Gogol, Goa

(b) The offerings for Rakhandar, including maadi, rice bread, beedi, collected before the ritual.

Fig. 2 Rakhandar temple

3.2 Primary Research

The primary research for this study involved firsthand accounts gathered through personal interactions with individuals who had either heard about Rakhandar or had direct experiences related to this fascinating entity. Additionally, visits were made to the Rakhandar shrines. Through these interactions and observations, a wide range of narratives surrounding Rakhandar worship were collected, encompassing both chilling and humorous anecdotes. The inclusion of firsthand accounts in this research contributes to the richness and authenticity of the findings. The interviews were informal, and the participants were asked about any experiences they had with Rakhandar folklore. The goal was to get interesting stories surrounding this entity, which could further shape my narrative (Fig. 2).

4 Analysis

The research culminated in a diverse array of perspectives from various sources, encompassing opinions on people, texts, music, and more. Rakhandar was widely interpreted as benevolent, mischievous, strict, and, in some cases, even malevolent. By leveraging these varied viewpoints, we were able to craft distinct narratives about Rakhandar as relayed by different village members. This approach allowed us to highlight the villagers' reliance on Rakhandar, whether it be for prosperity or protection. Given Rakhandar's pre-Brahmanic roots, distinct offerings were required, diverging from the customary practices associated with mainstream Hindu Brahmanic deities. The narrative would centre on these offerings, each quest to claim these ritual items unveiling different flavours of character interactions (Fig. 3).

Fig. 3 Screenshots from the game. Top left: The village fisherwoman interacting with the main character. Top right: Sarvesh attempting to tap Feni through the distillery. Bottom: The village matron narrating her experience with Rakhandar

5 Outcome

The outcome of this project is a digital game. It vividly depicts the Konkani folklore of Rakhandar from the viewpoint of Sarvesh, a fourteen-year-old city dweller who visits his maternal hometown. Numerous interactions with the villagers reveal the rich culture that surrounds Rakhandar. These interactions together create a story that the user discovers as it develops. The opposing perceptions of Rakhandar—his generous, forgiving side and his imposing persona—emerge across two unique game levels, night and day.

5.1 Walkthrough

The game commences with an intriguing cutscene that interweaves world-building, character introductions, and intriguing elements to captivate the player's attention. Developed within the Unity game engine, the cutscene leverages the timeline feature and different camera angle choices to elevate the storytelling experience (Fig. 4a). Onboarding elements such as the 'run' button are embedded in-game at strategic points throughout the gameplay to ensure that users are familiarized with basic interactions. These elements are carefully implemented to avoid disrupting the game's immersive experience. Multiple interaction points are labelled with the appropriate UI button to guide users in engaging with in-game characters and elements (Fig. 4b).

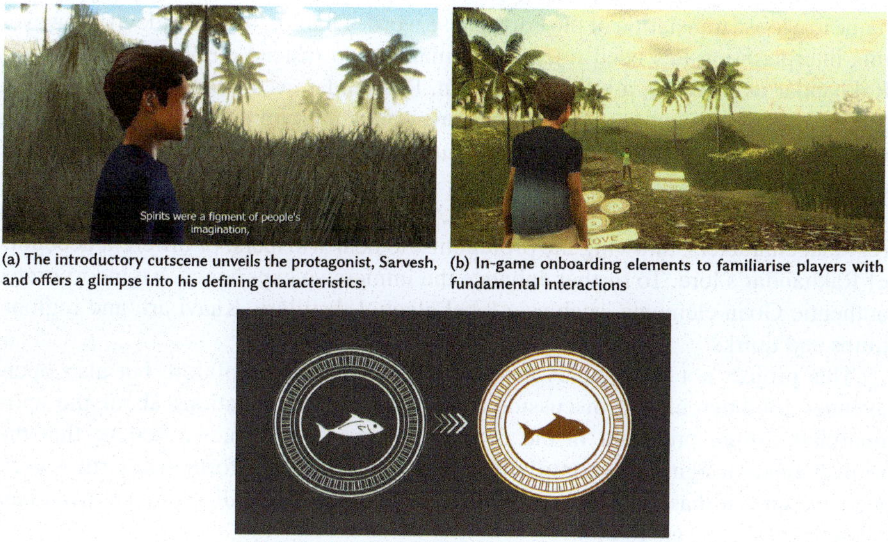

(a) The introductory cutscene unveils the protagonist, Sarvesh, and offers a glimpse into his defining characteristics.

(b) In-game onboarding elements to familiarise players with fundamental interactions

(c) Example of a UI element that updates to a Kaavi art-inspired filled state when the user completes a task

Fig. 4 In-game cutscenes and UI elements

A list of objectives is also added in the form of UI elements. All in-game UI elements were designed with Kaavi art as an inspiration. Kaavi art is a form of murals found in the Konkan region, especially in temples of Goa, parts of coastal Maharashtra, and Karnataka. Kaavi murals can also be seen in old houses and small shrines. Kaavi art uses a deep red pigment, which is reflected in the design of the UI elements. As the gamer completes their tasks, the UI element changes to a filled state, which is coloured in the same way as a completed Kaavi art mural (Fig. 4c).

6 Conclusion

In conclusion, this study delves into the complex folklore surrounding Rakhandar, a spirit with a pervasive presence in the Konkan region. Through a comprehensive examination of diverse sources, including books, audio recordings, and videos, we were able to map the multifaceted understanding of Rakhandar's role and influence.

Rakhandar's duality in perception emerges as a striking feature of the folklore. While revered and adored by many, he is simultaneously viewed with a tinge of fear and apprehension. This intriguing contrast stems from Rakhandar's dual nature as a spirit god, embodying both protective and fearful aspects. We explored the multiple origin stories associated with Rakhandar, with one notable interpretation linking his name to a loose translation of the Portuguese word "Diablo," meaning devil. However, this interpretation has been rejected by many, who instead embrace the notion of Rakhandar as Devchar, a village guardian. Rakhandar, as a guardian, symbolizes a profound commitment to safeguarding, preserving, and nurturing both the land and the encompassing ecology. This call for cultural and ecological preservation assumes a central role in our constructed narrative.

Leveraging this duality of perception, we crafted a virtual world teeming with regional characters, allowing the protagonist to hear whispers, or different accounts, of Rakhandar's lore. To further enhance the immersive experience, we incorporated authentic Goan elements, such as a local alcohol distillery, Kaavi art, and regional quips and quirks.

This project not only yields insights into Rakhandar's folklore but also opens avenues for intercultural discussions. By sparking conversations about the spirit guardian culture prevalent in the coastal region of Konkan, we believe that this project makes a significant contribution. The engaging and informative virtual world we have created has the potential to educate, entertain, and disseminate awareness to audiences for years to come.

7 Future Scope

To enhance player engagement and interactivity, the game can be further developed to incorporate more control over character conversations and branching narratives. By granting players greater autonomy within the gameplay, their involvement and interest in the game's progression will be amplified.

Additionally, introducing a wider range of interactive elements will transform the game into a more immersive and engaging experience. By incorporating challenging yet enjoyable tasks and collectables, players will be driven to explore the game's world more comprehensively, fostering a sense of accomplishment and motivation. Furthermore, to create a greater sense of Goan belonging, dedicated models of buildings, flora, and fauna can be added. More scope for exploration, hidden incentives, and easter eggs can be incorporated to polish this game further.

India's folklore offers a treasure trove of compelling stories that transcend the conventional good versus evil dichotomy. These narratives, often imbued with moral complexities and grey undertones, hold immense potential for captivating storytelling. As a future scope, diverse tales from various regions of India, such as the captivating lore of Buta Kola from the Tulu Nadu region, popularized by the movie "Kantara," and many other similar tales on the themes of protector spirit-gods, could be documented and re-imagined through various mediums.

References

1. Dutta U (2019) Digital preservation of indigenous culture and narratives from the Global South: in search of an approach. Humanities 8(2). https://doi.org/10.3390/h8020068
2. Eagles A, Gallant K (2017) First-person vs. third-person. https://api.semanticscholar.org/CorpusID:148692787
3. Gomez A (2015) Case of the missing Rakhondar. Accessed 12 May 2023. https://www.thegoan.net//case-of-the-missing-rakhondar/4503.html
4. Khanna R, Bhairav JF (2020) Ghosts, monsters and demons of India. Blaft Publications Pvt. Ltd., Inc
5. Panandiker SP (n.d.) Rakhno—the village protector. Talking myths—an online archive of traditional tales from Indian subcontinent. Accessed 12 May 2023. https://talkingmyths.com/rakhno-the-village-protector/
6. Shliakhovchuk E, Muñoz A (2020) Intercultural perspective on impact of video games on players: insights from a systematic review of recent literature. Educ Sci Theory Pract 20:40–58. https://doi.org/10.12738/jestp.2020.1.004
7. Stegner B (2020) First-person games vs. third-person games: what are the differences? Accessed 16 Oct 2023. https://www.makeuseof.com/first-person-games-vs-third-person-games-differences/
8. United Cities and Local Governments (2018) Culture in the sustainable development goals (SDGs): a guide for local action. United Cities and Local Governments. https://www.uclg.org/sites/default/files/culture_in_the_sdgs.pdf

KaavadBits: Interactive Art Installation for a New Narrative Experience

Saumik Shashwat, Aditya Padmagirwar, Shivoy Arora, and Anmol Srivastava

Abstract This work explores branching narratives through *KaavadBits*, a table-top art installation embodying the *kaavadiya-jajman* (narrator-patron) perspective of *kaavad baanchana*, the Indian storytelling tradition of reciting the *kaavad*. For diegetic worldbuilding of tales from *Panchatantra*, a compilation of ancient Indian animal fables, the narrator takes the physical form of a tree, with which the audience interacts using tokens for a seamless multi-modal storytelling experience. Building on the related explorations, we propose a design that immerses the audience through choice, character and question-based interactions, followed by insights from its user study. Through this paper, we aim to strike consequential tangible, technological and narrative explorations into various lesser-known traditional forms of storytelling that may inspire new interaction techniques, ultimately preserving the art form.

Keywords Embodied interaction · Tangible interaction · Interactive storytelling · Branching narratives · Kaavad

---

Saumik Shashwat and Aditya Padmagirwar and Shivoy Arora contributed equally to this research.

S. Shashwat (✉) · A. Padmagirwar · S. Arora · A. Srivastava
Creative Interfaces Lab, Department of HCD, IIIT Delhi, New Delhi, India
e-mail: saumik20404@iiitd.ac.in

A. Padmagirwar
e-mail: aditya20449@iiitd.ac.in

S. Arora
e-mail: shivoy21420@iiitd.ac.in

A. Srivastava
e-mail: anmol@iiitd.ac.in

A. Joshi and G. R. Sim (eds.), *Proceedings of the 14th Indian Conference on Human-Computer Interaction*, Studies in Computational Intelligence 1162,
https://doi.org/10.1007/978-981-97-4335-3_11

CCS Concepts:• **Human-centered computing → Interaction design; Human computer interaction (HCI)**

1 Introduction

> Story is perhaps the oldest form of communication known to humankind. It has a way of mesmerising the listeners into silence and the tellers into expressing the deepest desires and anxieties of their society, directly or through subversive means.—Nina Sabnani in 'The Kaavad storytelling tradition of Rajasthan'. [1]

Storytelling and its practice have taken many forms over the centuries: voices, gestures, dance, music, performances, or their combination, brought people together and shaped cultures. India has a rich heritage in its diversity of storytelling forms, passing countless intricate tales over generations. The tradition of listening to stories from the great epics is considered sacred by many believers. Such is the power of stories and storytelling traditions. One such is the approximately 400-year-old oral tradition of *kaavad baanchana* from Rajasthan in India [2].

Kaavad (Fig. 1) is a portable wooden shrine featuring illustrated narratives on its panels that opens like *kivads* (doors), hinged together which open and close to reveal its intricately painted images of epics and other local/folk stories. These stories could be religious (Hindu epics like Ramayana and Mahabharata), mythical, historical (local legends/folk tales) or from the present (depicting the daily struggles of people) that the *kaavadiya* (*kaavad* storyteller) weaves into beautiful narrations [2]. The experience is multi-sensory as the storyteller uses paintings, narration and gestures to tell the stories. Storytellers collaborate with visual artists to personally create imagery to act as mnemonics for the events from the stories, and often the same images to tell different stories each time. But today, this unique storytelling tradition is diminishing [2].

Using *Kaavad* as an anchor, we employed a research-through-design (RtD) [4] to create an initial prototype for *KaavadBits* to emulate close to the *kaavad banachana* experience as a new media experience. Our rapid literature review on interactive storytelling helped us identify some gaps in research on branching narratives in tangible interactive storytelling and under-representation of non-western contexts, mainly traditional Indian storytelling forms. Some of the notable works that inspired us are [2, 5–9]. The name *KaavadBits* explains the binary-choice-based storyline progression and the tangible tokens of *kaavad*, and it also pays homage to the breakthrough work of Hiroshi Ishii and Brygg Ullmer on Tangible Bits [10].

Fig. 1 *Kaavad*, the Rajasthani storytelling tradition, depicts the illustration of its stories on the doors [3]

2 Design

2.1 Methodology

We realised the design for *KaavadBits* over multiple iterations involving sketches and cardboard prototypes to visualise the interactions. Our design motivation aimed to redirect and diverge the locus of attention from a 2D surface to a 3D interactive space. We collaboratively developed the installation using various rapid prototyping and DIY techniques, including laser-cutting MDF, NFC stickers, programmable LEDs, servo motors, Arduino Uno and Raspberry Pi 4.

2.2 Functional Components

(1) ***Kaavadiya* as a tree**: Portraying the *kaavadiya* as a tree integrates the narrator into the story world as most *Panchatantra* [11] tales are set in a forest, allowing for diegetic interactions. The tree features a servo motor-enabled oscillating canopy that directs user attention to the narrator when narrating. Branches leading up to the canopy are represented by an LED matrix illuminating the chosen narrative path. The canopy houses special figurines for diegetic interactions- apple, mango, pear, and 'The Bloom of Eternal Tales'. Two hanging roots protruding from the canopy enable the user to control the branching storyline progression (Fig. 2a).

(2) **NFC-enabled Story blocks and Figurines**: Traditionally, the *kivads* are illustrated with pictorial descriptions of the stories, for which we reimagined the use-case to represent whole stories and their characters- there are four NFC-enabled laser-cut MDF story blocks with artworks we generated using Playground AI

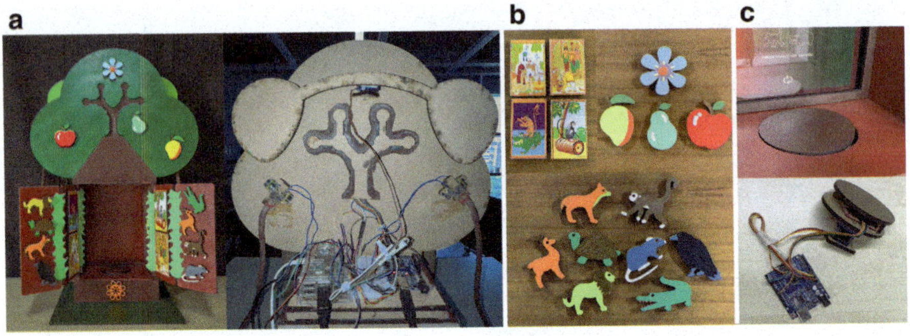

Fig. 2 **a** Front and back sides of the installation. **b** Story blocks and Figurines. **c** Stage area and Central platform

[12] and eight hand-painted character figurines which represent animals from *Panchatantra* (Fig. 2b).

(3) **Stage area and Central platform**: We reimagined the traditionally sacred central cavity [1] as the storytelling stage area with a display backdrop featuring corresponding images, captions, and interaction prompts. An Apple iPad Air device that can be slid into the slot is used as the display and is connected to the Raspberry Pi using VNC. A central platform featuring an NFC scanner for scanning sticker UIDs oscillates when the placed character is contextually central to the point in the story, such as when delivering a dialogue (Fig. 2c).

3 Interactive Narratives

Panchatantra tales carry valuable lessons about friendship, trust, and the consequences of our choices. We sourced our storylines from [11] and broke it down into a story tree [13] and prompted ChatGPT running GPT3.5 [14] to generate the alternative endings. The raw outputs were manually fine-tuned by us to remove the inconsistencies and set the tone descriptively closer to that of *kaavad-baanchna*. We self-recorded the narrations as the open-source text-to-speech models yielded robotic-sounding speeches and had time and language limitations. The recorded clips were then arranged and programmed to their respective points in the story. The user interaction task flow is as follows:

(1) **Onboarding**: User interaction with *KaavadBits* starts with them opening the *kivads* as they get acquainted with the characters and the stories, revealing the stage area. The narrator greets the audience and asks them to choose and present a story card and then its characters. Upon receiving the patronage in the form of 'The Bloom of Eternal Tales', the narrator begins enacting the story.

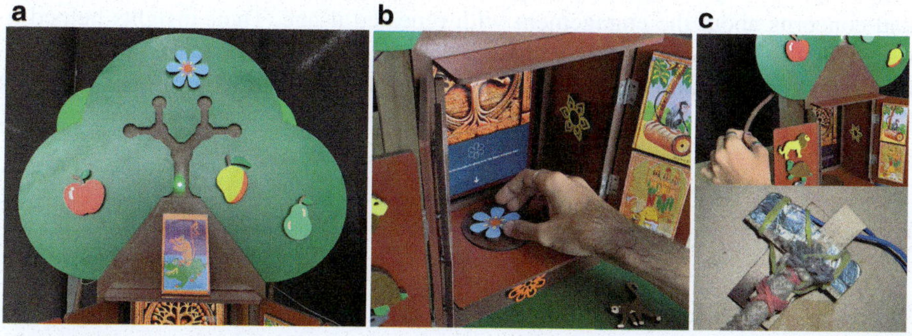

Fig. 3 **a** First LED lighting up after onboarding is complete. **b** The user is placing the figurine on the central platform. **c** The User pulls the left branch at a diverging point in the story and its mechanism

(2) **Figurine Associations**: Following onboarding, the user interacts with the character and special figurines throughout the story by replacing the figurine on the central platform. This interaction serves a dual purpose- immersing the audience by giving them an 'agency' and directing their attention to the points in the story.
(3) **Choice-based branches**: At specific diverging points in a storyline, the user is presented with a binary choice that directs the rest of the story and its ending. As our current artefact constitutes two such instances, stories on KaavadBits have four unique endings.
(4) **Persona Questions**: After the moral of an ending is narrated, three special tokens are presented to the audience- an apple to question the narrator, a mango and a pear to ask the two primary characters. We realised this interaction using Wizard of Oz prototyping for our user evaluation (Fig. 3) [15].

4 User Evaluation and Findings

Our primary objectives were to investigate how people interacted and felt about a diegetic tangible user interface and its efficacy in engaging the audience. Eight participants were sampled from our university's student population. Each 30-minute session in our research lab included a project briefing, pre-test discussion, exploratory session with *KaavadBits*, system-usability scale testing [16], and post-test discussion. The semi-structured discussions covered participants' awareness of Indian storytelling traditions, interactive storytelling mediums, and their experience interacting with the installation. We also did a System Usability Scale (SUS) [16] to judge the system's overall usability and user experience. The system reported a SUS score of **78.75**, which is well above the baseline score for a usable system. Qualitative interviews with the preliminary users made us aware of a significant shortage of knowledge about the *kaavad* art form. Though the participants liked the overall experience, they had concerns about the engagement with repeated usage. Their insights helped us realise that future work can build upon creating a learning experience from this system, which can be more appropriately engaging for younger children for repetition learning.

5 Conclusion and Future Work

This paper introduced a tangible interactive storytelling art installation inspired by the tradition of *Kaavad* for tales from *Panchatantra*. In its future iterations, this system can take an abstract form that can combine stories from multiple universes as demanded by its user, with automated story creation pipelines that use large-language models to create structured storylines. Future work also aims to augment more novel interactions inspired by non-Western fiction. We also aim to test the installation further with more users and get more comprehensive insights to direct our next iteration.

References

1. Nina Sabnani. 2009. The Kaavad storytelling tradition of Rajasthan. (2009)
2. Nina Sabnani. 2007. Sharing Stories: Learning with Stories. In *Usability and Internationalization. Global and Local User Interfaces* (Lecture Notes in Computer Science), Nuray Aykin (Ed.). Springer, Berlin, Heidelberg, 460–468. https://doi.org/10.1007/978-3-540-73289-1_53
3. Hiart. 2014. English: Kavad from Rajasthan, early 20th century, paint on wood, maker unknown. https://commons.wikimedia.org/wiki/File:Kavad_from_Rajasthan,_early_20th_century,_paint_on_wood,_maker_unknown.JPG
4. Frayling C (1994) Research in art and design. Royal Coll Art Res Pap 1(1):9. https://researchonline.rca.ac.uk/384/. Royal College of Art, London
5. Chenzira A, Chen Y, Mazalek A (2008) RENATI: recontextualizing narratives for tangible interfaces. In: Proceedings of the 2nd international conference on Tangible and embedded interaction (TEI'08). Association for Computing Machinery, New York, NY, USA, pp 147–148. https://doi.org/10.1145/1347390.1347423
6. Chopra B, Gupta R (2022) StoryBox: independent multi-modal interactive storytelling for children with visual impairment. In: Extended abstracts of the 2022 CHI conference on human factors in computing systems (CHI EA'22). Association for Computing Machinery, New York, NY, USA, pp 1–7. https://doi.org/10.1145/3491101.3519651
7. Chu JH, Clifton P, Harley D, Pavao J, Mazalek A (2015) Mapping place: supporting cultural learning through a Lukasa-inspired tangible tabletop museum exhibit. In: Proceedings of the ninth international conference on tangible, embedded, and embodied interaction (TEI '15). Association for Computing Machinery, New York, NY, USA, pp 261–268. https://doi.org/10.1145/2677199.2680559
8. Gupta S, Tanenbaum TJ, Tanenbaum K (2019) Shiva's rangoli: tangible storytelling through diegetic interfaces in ambient environments. In: Proceedings of the thirteenth international conference on tangible, embedded, and embodied interaction (TEI '19). Association for Computing Machinery, New York, NY, USA, pp 65–75. https://doi.org/10.1145/3294109.3295635

9. Lizette Reitsma, Andrew Smith, and Elise van den Hoven. 2013. StoryBeads: Preserving Indigenous Knowledge through Tangible Interaction Design. In *2013 International Conference on Culture and Computing*. 79–85. https://doi.org/10.1109/CultureComputing.2013.22
10. Hiroshi Ishii and Brygg Ullmer. 1997. Tangible bits: towards seamless interfaces between people, bits and atoms. In *Proceedings of the ACM SIGCHI Conference on Human factors in computing systems(CHI '97)*. Association for Computing Machinery, New York, NY, USA, 234–241. https://doi.org/10.1145/258549.258715
11. 2023. Panchatantra Online Stories. http://panchatantra.org/panchatantra-stories.html
12. Playground AI (2023) Free AI image generator: art, social media, marketing. https://playgroundai.com/
13. Alfieri L (2021) The past, present & future of interactive narrative storytelling: multi-branching narratives. https://blooloop.com/theme-park/opinion/multi-branching-narrative/
14. OpenAI. 2023. ChatGPT. https://chat.openai.com
15. Dahlbäck N, Jönsson A, Ahrenberg L (1993) Wizard of Oz studies - why and how. Knowl Based Syst 6(4):258–266. https://doi.org/10.1016/0950-7051(93)90017-N
16. Assistant Secretary for Public Affairs (2013) System usability scale (SUS). https://www.usability.gov/how-to-and-tools/methods/system-usability-scale.html. Department of Health and Human Services

Jail Diaries: An Interactive Pop-Up Book Art Installation

Biswasmita Mishra, Aishiki Bhattacharya, Ankush Gupta, Siddharth Gupta, and Anmol Srivastava

Abstract In this paper, we introduce an interactive pop-up book installation called 'Jail Diaries.' The book installation is designed for step-by-step interactions, including sound, touch, and visuals, with the goal of enhancing the storytelling process. Jail Diaries aims to connect users with the history of Indian freedom fighters and their sacrifices for the country by using technology to bring a pop-up book to life.

Keywords Interactive storytelling · Pop-up book · Interactive narrative · Interactive ambience

CCS Concepts:• **Human-centered computing → Interaction devices; Interaction design**

B. Mishra (✉) · A. Bhattacharya · A. Gupta · S. Gupta · A. Srivastava
Indraprastha Institute of Information Technology, Okhla, Delhi 110020, India
e-mail: biswasmita21317@iiitd.ac.in

A. Bhattacharya
e-mail: aishiki21007@iiitd.ac.in

A. Gupta
e-mail: ankush21232@iiitd.ac.in

S. Gupta
e-mail: siddharth21355@iiitd.ac.in

A. Srivastava
e-mail: anmol@iiitd.ac.in

A. Joshi and G. R. Sim (eds.), *Proceedings of the 14th Indian Conference on Human-Computer Interaction*, Studies in Computational Intelligence 1162,
https://doi.org/10.1007/978-981-97-4335-3_12

1 Introduction

In today's era, ebooks and digital reading platforms have revolutionized how people engage with literature. However, it is essential to acknowledge that physical books with pop-up features still hold a special place in the reading experience due to their ability to provide tactile and emotional involvement. Previously work has been done on material experimentation, artistic design, and engineering for pop-up books [3].

This paper aims to shed light on our project centered on the themes of Disobedience, Defiance and Revolution. Blending various elements such as 2D pop-ups, coordinated lights and audio, it focuses on the stories of four freedom fighters from India who fought for the rights of the people and were imprisoned for their acts of disobedience. These individuals were brave, as they stood up against the powerful government of their time to fight for media rights, women's rights, jail rights, and other important causes that are now widely accepted. The project also involves elements such as an interactive light, audio associated with each page that can be played voluntarily and a projection controlled by the movement of the pages. Such elements immerse the viewers in the experiences of the freedom fighters, transporting them back in time to witness key scenes from their lives, creating an ambiance that captures the oppressive atmosphere of the colonial era while also highlighting the indomitable spirit and hope that fueled the revolutionaries. Our book serves as a diary offering a narrative perspective that delves deeper into the thoughts, emotions, and challenges freedom fighters face. We aim to create a connection with the readers by appealing to their senses. The tactile nature of our book encourages readers to interact with the pages, turning each moment into a rich multi-sensory journey. By touching and experiencing the textures and pop-up elements, readers can deepen their bond with the story, while listening to the narrated voices adds an even more profound and lasting impact.

2 Background

2.1 History—Relevance in Today's World

This project holds immense importance in today's world because it educates, inspires, and raises awareness about the struggles of Indian freedom fighters. The project aids in our understanding of the underlying reasons of many societal difficulties that still exist in our society today by examining the lives of these brave people. These activists campaigned for essential rights that are still important today, such as women's rights, media freedom, and better prison conditions. Visitors can make links between the past and the present by learning about their stories, which will help them better comprehend the difficulties that our society is currently facing. Users are inspired to make positive changes in their own communities as a result of witnessing the bravery of these heroes. The harsh conditions these fighters endured in colonial prisons evoke empathy, highlighting the urgent need to defend the rights and dignity of every individual, regardless of their background. These prison diaries also have a huge impact on India's cultural past. For future generations to stay connected to their past, these narratives must be preserved. Young people are motivated to cherish and safeguard their cultural inheritance when they learn about the sacrifices made by their predecessors. These tales act as a lighthouse, inspiring people to defend what is pure and right. By doing this, they promote a sense of pride, cultural identification, and empathy-all of which are crucial for ensuring a brighter future for our society.

2.2 *How an Installation Like This Can Help in Today's World*

The installation of an interactive pop-up book offers a sweet spot between digital and physical media consumption. Today, information is available abundantly, so how it is presented has become much more valuable and it determines how well it would be understood and assimilated by the users. Thus this installation also aims at using its unique format to minimize the effort required to gain knowledge. Such a pop-up book installation will add interest to the education process for children, and thus can be made use of in institutes like schools and in museums. Previous research has explored tangible storytelling installations, including [2] which empowers readers to shape narrative emotions via their physical space. It shows how such interfaces foster reader engagement and co-authorship, blurring real and fictional boundaries.

2.3 *Pop Up Book—The Visual Impact*

The detailed pop-ups bring to life the characters and their experiences and transports the users into that time and era. The users are able to unfold scenes from the time of the Independence movement before their eyes and understand and relate to the brave freedom fighters on a personal level. The book also has unique textures and audio clips that support the visuals and enable a more immersive experience for the users and allow for a deeper connection with the story. Physically interacting with the stories of these freedom fighterâŁ™s lives and their commitment to social justice and freedom could inspire future generations of activists and serve as a call to action for the users to work for a more just and equitable world. Previous works have explored interactive digital storytelling for children by using a digital pop-up book where they can search for the story (Fig. 1) [1].

3 Installation

3.1 *Installation Requirements*

A pop-up book, a book stand, a lamp, a speaker and a projector setup are all part of the installation. We have used multiple types of paper material for the pop-up book.

Fig. 1 Pop-up book

Four key pieces of hardware are required to build this setup: a lamp, a projector, an Arduino board, conductive ink, as well as some resistors and connecting cables.

In this installation, a pressure plate on a chair in a dark room triggers a light bulb, illuminating the space when someone sits on the chair. The user's involvement causes a projector to simultaneously show the pop-up book's contents on a screen. Specific spots marked with conductive ink can be found on the book's pop-up pages. When the user touches any of these spots, pre-programmed code on an Arduino board connected to the software, Processing, is run. Pre-recorded sounds are played back when this code is activated. Additionally, the presentation on the screen changes as the user turns the pages of the book, ensuring a seamless encounter between the digital wall projection and the actual book.

Using laser cutting methods, the installation's aesthetic appeal and user-friendliness were improved. A book stand and distinctive lighting components were made using laser cutting. These improvements not only enhance the lighting setup but also deftly hide the projector inside the book stand, maintaining the installation's immersive atmosphere. These elements and technology work together to produce a seamless connection between the user and the artwork that is both engaging and intriguing.

3.2 Dark Room, Silent Environment (Situation)

The user begins the exhibit by sitting on a chair in a dark room, meant to make them feel what it is like being inside a jail. In the middle of the setup, there's one chair,

Fig. 2 Installation Setup

and when they sit down, a lamp above them lights up as a surprise, shining over the pop-up book in front. It's quiet all around. When they open the book, each page has a special ink spot which, when touched, triggers an audio clip explaining that page's topic and some narration from the concerned freedom fighter's life story. We made these sounds to leave a big impact on the users. The sound script is carefully made, with parts of the Freedom Fighter's daily struggles included. When they sit down and start turning the book's pages, a projector changes what it shows to match the page's topic. When they finish the book and get up, the lights go off, indicating that the storytelling is done (Fig. 2).

4 Future Prospects

We anticipate a multitude of exciting opportunities and potential growth of our project. Collaborations with historical societies, museums, and art galleries could lead to traveling exhibitions, allowing the stories of Indian freedom fighters to reach broader audiences both nationally and internationally. Moreover through collaborations with institutions, government agencies and non-governmental organizations (NGOs) we have the opportunity to broaden our influence in schools. This will not only enhance history education but also serve as a source of inspiration for the upcoming wave of individuals who will drive positive change in society. Additionally, the prospect of integration into Indian Revolution-based theme parks presents the opportunity to create dynamic, immersive attractions that delve deeper into the lives and struggles of these historical figures.

5 Conclusion

In conclusion, our project renews the appreciation for physical pop-up books as vessels of history, bridging generations with the powerful narratives of Defiance and Revolution. It serves as a testament to the enduring allure of tangible storytelling in a digital age.

References

1. Ahooja L (2022) Edutainment with a Mixed reality pop-up book for children: Augmenting digital interventions in the popup book. In: Proceedings of the 12th Indian Conference on Human-Computer Interaction (IndiaHCI'21). Association for Computing Machinery, New York, NY, USA, pp 114–116. https://doi.org/10.1145/3506469.3506488
2. Gupta S, Tanenbaum TJ, Tanenbaum K (2019) Shiva's rangoli: tangible storytelling through diegetic interfaces in ambient environments. In: Proceedings of the thirteenth international conference on tangible, embedded, and embodied interaction (TEI'19). Association for Computing Machinery, New York, NY, USA, pp 65–75. https://doi.org/10.1145/3294109.3295635
3. Qi J, Buechley L (2010) Electronic popables. In: Proceedings of the fourth international conference on tangible, embedded, and embodied interaction. https://dl.acm.org/doi/10.1145/1709886.1709909

SPRINGER NATURE

GPSR Compliance

The European Union's (EU) General Product Safety Regulation (GPSR) is a set of rules that requires consumer products to be safe and our obligations to ensure this.

If you have any concerns about our products, you can contact us on ProductSafety@springernature.com

In case Publisher is established outside the EU, the EU authorized representative is:

Springer Nature Customer Service Center GmbH
Europaplatz 3
69115 Heidelberg, Germany

The manufacturer's authorised representative in the EU is Springer Nature Customer Service Centre GmbH, Europaplatz 3, 69115 Heidelberg, Germany. If you have any concerns regarding our products, please contact ProductSafety@springernature.com

Printed and bound by CPI Group (UK) Ltd, Croydon, CR0 4YY
01/10/2026
02225919-0019